Traffic Simulation
and Data

Validation Methods and Applications

Traffic Simulation
and Data

Validation Methods and Applications

Edited by
WINNIE DAAMEN
CHRISTINE BUISSON
SERGE P. HOOGENDOORN

CRC Press
Taylor & Francis Group
Boca Raton London New York

CRC Press is an imprint of the
Taylor & Francis Group, an **informa** business

CRC Press
Taylor & Francis Group
6000 Broken Sound Parkway NW, Suite 300
Boca Raton, FL 33487-2742

First issued in paperback 2017

© 2015 by Taylor & Francis Group, LLC
CRC Press is an imprint of Taylor & Francis Group, an Informa business

No claim to original U.S. Government works
Version Date: 20140320

ISBN 13: 978-1-138-07591-7 (pbk)
ISBN 13: 978-1-4822-2870-0 (hbk)

Visit the Taylor & Francis Web site at
http://www.taylorandfrancis.com

and the CRC Press Web site at
http://www.crcpress.com

Contents

7 Validation 163

CONSTANTINOS ANTONIOU, JORDI CASAS, HARIS KOUTSOPOULOS, AND RONGHUI LIU

8 Conclusions 185

CHRISTINE BUISSON, SERGE HOOGENDOORN, AND WINNIE DAAMEN

List of Figures

List of Tables

Preface

This book has been written within work package 1 of the Multitude project. Multitude was a European network funded by COST, an intergovernmental framework for European Cooperation in Science and Technology. The main objective of Multitude was to develop, implement, and promote the use of methods and procedures for supporting the use of traffic simulation models, especially regarding model calibration and validation, to ensure their proper use and the validity of the results and the decisions made on them. In order to reach the objective, four work packages were defined:

1. State of the art of traffic simulation practice and research
2. Highway modelling
3. Network modelling
4. Synthesis, dissemination and training

This book is one of the three deliverables of work package 1. The other two deliverables are a report on the state of the practice and an overview of national guidelines on performing traffic simulations and calibration and validation of traffic models.

We would like to express our thanks to Mark Brackstone (IOMI, United Kingdom), Arie van Ekeren (Delft University of Technology, the Netherlands), Victor Knoop (Delft University of Technology, the Netherlands), Pete Sykes (SIAS, United Kingdom), Tom van Vuren (Mott MacDonald, United Kingdom). Without your contributions and those of the authors, it would have been impossible to deliver this state-of-the-art document.

The Editors

List of Contributors

Constantinos Antoniou
National Technical University
of Athens
Athens, Greece

Jean-Michel Auberlet
IFSTTAR
Lyon-Bron, France

Carlos Azevedo
Laboratório Nacional de
Engenharia Civil
Lisbon, Portugal

Jaume Barcelo
Universitat Politècnica de Catalunya
Barcelona, Spain

Ashish Bhaskar
Queensland University of
Technology
Brisbane, Australia

Christine Buisson
IFSTTAR
Lyon-Bron, France

Jordi Casas
TSS
Barcelona, Spain

Biagio Ciuffo
Joint Research Centre
European Commission
Ispra, Italy

Winnie Daamen
Delft University of Technology
Delft, the Netherlands

Haneen Farah
Delft University of Technology
Delft, the Netherlands

Gunnar Flötteröd
Kungliga Tekniska Högskolanm
Stockholm, Sweden

Raymond Hoogendoorn
Delft University of Technology
Delft, the Netherlands

Serge Hoogendoorn
Delft University of Technology
Delft, the Netherlands

Tanya Kolechkina
Technion – Israel Institute of
Technology
Haifa, Israel

Haris Koutsopoulos
Kungliga Tekniska Högskolanm
Stockholm, Sweden

Axel Leonhardt
PTV AG
Karlsruhe, Germany

Hans van Lint
Delft University of Technology
Delft, the Netherlands

Ronghui Liu
University of Leeds
Leeds, England

Vittorio Marzano
Università di Napoli Federico II
Naples, Italy

Marcello Montanino
Università di Napoli Federico II
Naples, Italy

Vincenzo Punzo
Università di Napoli Federico II
Naples, Italy

Thomas Schreiter
University of California at Berkeley
Berkeley, California

Tomer Toledo
Technion–Israel Institute of
   Technology
Haifa, Israel

Peter Wagner
Forschungszentrum der
   Bundesrepublik Deutschland für
   Luft- und Raumfahrt
Köln, Germany

Yufei Yuan
Delft University of Technology
Delft, the Netherlands

Chapter 1

Introduction

Winnie Daamen, Christine Buisson,
and Serge Hoogendoorn

Traffic and transportation applications are rapidly expanding in scope due to their potential impacts on community and environmental decision making. These applications range from planning and assessment of road infrastructure to evaluation of advanced traffic management and information systems (e.g., dynamic hard-shoulder running) and testing technologies and systems to increase safety, capacity, and environmental efficiency of vehicles and roads (e.g., cooperative systems and intelligent speed adaptation). The complexity and scale of these problems dictate that accurate and dynamic traffic simulation models rather than analytical methods are used increasingly for these purposes.

Many commercial traffic simulation models are currently available, and even more models have been developed by research institutes and research groups all over the world. However, the simulation results should be interpreted with great care. First, the quality of the simulation models should be considered. In addition, the reproducibility of the simulation results is important. Reproducibility is the ability of simulation results to be accurately reproduced or replicated by a party working independently using the same or a different simulation model. Since more and more parameters must be set in traffic simulation models, situations can be modeled in different ways and models exhibit increasing complexity, the capabilities of a user may affect the quality of the simulation results.

Therefore, it is important to develop methods and procedures to help developers and users to apply traffic simulation models correctly, effectively, and with reproducible results. Motivations and solutions to this problem should be found in the traffic models themselves and in the ways they are applied, following an approach that is often halfway between deductive and inductive, "whereby one first develops (via physical reasoning and/or adequate idealizations and/or physical analogies) a basic mathematical modeling structure and then fits this specific structure (its parameters) to real data" (Papageorgiou, 1998). The fitting process is generally known as model calibration. Validation tests whether a model gives a sufficiently accurate representation of reality (Kleijnen, 1995). As for calibration, during the validation of a simulation tool, predictions from the simulation model are compared to observations from reality, but a data set different from the data set used for calibration should be utilized.

Unfortunately, calibration and validation against suitable observed data are not commonly practiced in the field of traffic simulation. Until now, no standardized methods existed and most efforts and resources focused on model (and software) development.

While researchers recently started working on these topics, the efforts are fragmented, based on different data sets, and motivated by various applications. The problem is further complicated by geographic and cultural differences in attitudes toward driving, road design, and traffic regulations among different countries, resulting in considerable differences in driving behaviors and traffic operations.

The aim of the MULTITUDE project (2013) covering methods and tools for supporting the use, calibration, and validation of traffic simulation models is therefore to develop, implement, and promote the use of methods and procedures to support the use of traffic simulation models, especially in relation to model calibration and validation, to ensure their proper use and the validity of the results and decisions based on them.

Before development and implementation of methods and procedures for calibration and validation can be started, an overview should indicate the information that is currently available on these and related topics. This overview can be used to identify the blank spots in the research and also to provide researchers and practitioners who are new in the field an opportunity to be introduced to existing (theoretical) knowledge about the calibration and validation processes in general and in performed calibrations and validations of specific models in particular. The aims of this state-of-the-art report are to:

- Analyze data collection techniques and estimation methodologies for innovative traffic data, e.g., vehicle trajectory data.
- Consider data reduction and enhancement techniques for standards, i.e., commonly available traffic information such as point detector data.
- Provide an overview of calibration and validation principles.
- Review literature on estimation, calibration, and validation of traffic flow models and corresponding methodologies, including estimating and refining travel demand matrices using traffic data.

First, we will look at the relationship of a real system and a simulated system, as shown in Figure 1.1. As indicated earlier, validation intends to determine how well a simulation model replicates a real system. In calibration, the outputs of the simulation and the real system are also compared, but the parameters of the simulated system are optimized until the difference between both outputs is minimal or at least meets specific minimum

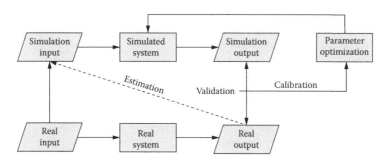

Figure 1.1 Relationship of simulated and real systems and locations of calibration and validation processes. (*Source:* Toledo, T. and Koutsopoulos, H. 2004. *Transportation Research Record*, 1876, 142–150. With permission.)

requirements. Ideally, the inputs of the real and simulated systems should be identical. Therefore, both the input variables and outputs of the real system should be observed. However, not all inputs (e.g., dynamic origin–destination matrices) can be observed directly and thus must be estimated; this introduces an additional source of inaccuracy.

A framework for calibration and validation of traffic simulation models is shown in Figure 1.2. Calibration and validation of traffic simulation models involve two steps (Toledo et al., 2003). Initially, the individual models of the simulator (e.g., driving behavior and route choices) are estimated using disaggregate data. Disaggregate data include detailed driver behavior issues such as vehicle trajectories. These individual models may be tested independently, for example, using a holdout sample. The disaggregate analysis is performed by statistical software and does not involve the use of a simulation model.

In the second step, the simulation model as a whole is calibrated and then validated using aggregate data (e.g., flows, speeds, occupancies, time headways, travel times, and queue lengths). Aggregate calibration and validation are important both in developing the model and applying it. The role of aggregate calibration is to ensure that the interactions of the individual models within the simulator are captured correctly and to refine previously estimated parameter values. In most practical applications, only aggregate traffic measurements are available. Model calibration in such cases must be performed by using aggregate data alone, so as to minimize the deviation between observed and simulated measurements.

Note, however, that the difference between aggregate and disaggregate data from the view of calibration is mostly a practical issue, not a fundamental one. Usually, disaggregate data are not available or are difficult to work with, but nothing forbids disaggregate testing of a simulation model.

This book starts with an overview of the various data collection techniques that can be applied to collect the different data types cited in Chapter 2.

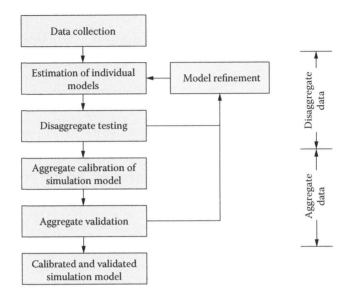

Figure 1.2 Calibration and validation framework of traffic simulation models. (*Source:* Toledo, T., Koutsopoulos, H.N., Davol, A. et al. 2003. *Transportation Research Record*, 1831, 65–75. With permission.)

Chapter 3 shows data processing and enhancement techniques for improving the quality of the collected data. The techniques are introduced according to the type of estimation, i.e., microscopic data enhancement, traffic state estimation, feature extraction and parameter identification techniques, and origin–destination (OD) matrix estimation. In Chapter 4, the principles of calibration and validation are described. In addition to generic procedures, the measures of performance, goodness of fit, and optimization algorithms are discussed.

Before focusing on the calibration and validation processes, Chapter 5 discusses the sensitivity analyses of the parameters in traffic models. These sensitivity analyses indicate the effects of various parameters on simulation results and thus on the importance of determining a correct value for a specific parameter. Chapter 6 gives details on network model calibration studies, while Chapter 7 focuses on the validation of simulation models. The final chapter discusses conclusions.

Chapter 2

Data collection techniques

Jean-Michel Auberlet, Ashish Bhaskar,
Biagio Ciuffo, Haneen Farah,
Raymond Hoogendoorn, and Axel Leonhardt

CONTENTS

The objective of this chapter is to provide an overview of traffic data collection that can and should be used for the calibration and validation of traffic simulation models. There are big differences in availability of data from different sources. Some types of data such as loop detector data are widely available and used. Some can be measured with additional effort, for example, travel time data from GPS probe vehicles. Some types such as trajectory data are available only in rare situations such as research projects.

This means that a simulation study carried out as part of a traffic engineering project, having a restricted budget, typically must rely on existing loop data or can at most utilize some GPS probe drives. The objective of calibration and validation in a traffic engineering project is mainly to check whether a model of a specific area replicates—at a desired level of detail—the macroscopic traffic conditions (flow, speed, travel time) for a certain traffic demand. Consequently, data for calibration and validation in traffic engineering projects typically need not to be microscopic.

Conversely, data generated with much more effort (e.g., trajectory data) are typically used by researchers to investigate driver behavior in general.

Analysis of driving behavior such as car following and lane changing requires highly detailed data to generate adequate insight into the traffic features to be modeled. These data are typically very expensive and/or laborious to acquire.

Sections 2.1 through 2.7 briefly describe the technical backgrounds of various data types and detection techniques and discuss typical availability and application areas. Section 2.8 draws conclusions about what data to use for specific purposes. An overview table included in Section 2.8.4 may be useful to get a quick view on the various sorts of data that may be used for the calibration of microscopic traffic simulation models.

In accordance with the primary focus of this book, this chapter provides only an overview of data collection. Extensive literature covering the techniques and their performance is available to the public through the World Wide Web.

An interesting point is the expected quality of the data. However, there is some ambiguity in existing studies because "performance of a data collection system" is a result of several factors (hardware and software used, sensor configuration, and environmental and traffic conditions). Therefore, this chapter will not answer questions like "What is the expected accuracy?" and in "What sensor is best to be used?". Specific studies describing detector features and boundary conditions are cited.

Errors in data exert impacts on the calibration of a simulation model and hence, on its results. This impact is twofold. First, a calibration step is needed before a simulation can be performed. In Chapter 4, we show that errors in measuring the variables that are compared with the simulation results impact the optimal parameters set for the calibration process. Second, any simulation tool uses measured (or enhanced or estimated; see Chapter 3) variables as inputs. Therefore, data measurement errors must be kept in mind when performing simulation studies. The reader is invited to consult the available documentation to gain knowledge of limits and error bounds of each type of detector.

2.1 MANUAL RECORDING

Manual recording is not exactly a data collection technology but may become necessary if automatic data collection is not feasible or fails to provide sufficient insight. Manual observations may be especially useful at intersections. The following data can typically be collected manually:

- Traffic volumes
- Turning volumes at junctions
- Delays at signals
- Queue lengths

2.2 LOCAL DETECTOR DATA

This section describes local detector data in detail. First, the data characteristics are described, and then an overview of the various detector types and relevant information is presented.

2.2.1 Data characteristics

Local detector data constitute traffic information collected at a single measurement point on a road. Data can be raw (single vehicle data) or aggregate (information recorded at time intervals, typically 1, 5, 15, or 60 minutes, and in rare cases intervals smaller than 1 minute) covering one or several lanes. Depending on the detector type, raw data collected may include:

- Vehicle presence (time points when it enters and/or leaves the detection zone)
- Vehicle speed
- Vehicle class (truck, bus, etc.)
- True or sensor-specific (e.g., magnetic) vehicle length

Aggregate data based on a specific time interval may show:

- Vehicle count, possibly per vehicle class
- Average vehicle speed, possibly per vehicle class (time mean speed)
- Variance in time mean speed
- Local occupancy (fraction of time when vehicle was present in the detection zone)
- Average time headways and variances of time headways

Local traffic data are the most widely available automatically collected traffic information available now and play a key role in most simulation studies. Such data can be used as input values and boundary conditions to derive demand and route split rates. On the other hand, simulation output can be compared to local traffic data to validate a simulation. Comparing time series of local speeds is a common way to calibrate a simulation model for analyzing congestion development. Automatic origin–destination (OD) matrix correction is another application of local detector data. More details about OD matrix estimation and correction are given in Section 6.2 of Chapter 6.

2.2.2 Detector types

Detectors can roughly be classified as intrusive and nonintrusive. Conventional intrusive traffic collection devices primarily consist of inductive loop detectors. These detectors must be cut into a road surface. This makes them usable only as permanent detectors, as they cannot be used for short data collection periods. Because they are embedded in pavements, intrusive detectors are costly to install and maintain because they require road closures. Furthermore, they deteriorate under the impact of traffic. Loop detectors are well advanced because the technology has been applied for several decades. Another advantage is that they are less prone to vandalism and theft than nonintrusive devices.

Nonintrusive detectors are not in direct contact with vehicles and are usually side firing or mounted overhead. They experience less wear and tear than intrusive pavement-based detectors. Because they are not embedded in road surfaces, they are easier to install and to replace, making them suitable choices for temporary installations. Among the many technologies available are radar, ultrasonic, and video cameras. Some are advanced technologies used in the field for years. Others are still under development or involved in field trials.

Minge et al. (2010) noted that volume and speed measurement performance with state-of-the-art nonintrusive detection technology (radar, video, laser, infrared, and magnetometer) is satisfying, but classification remains a weak point, especially if standardized classification schemes such as FHWA 13 of the Federal Highway Administration in the United States (FHWA, 2001) are applied.

2.2.2.1 Inductive loop detectors

Inductive loops consist of wire loops inside a road surface. The loops are fed electrically with a frequency between 40 and 100 kHz. Any metal objects in the inside area of the loops change the electric inductivity of the loops and can be measured by an electronic device. Vehicle presence is the basic information provided by a loop detector. If two loops are combined in a small distance (typically a few meters), the speed of a vehicle can be measured with good accuracy.

Inductive loops are by far the most common detectors for road traffic. They are used as single loops around signals to provide information for vehicle-actuated control and on freeways as double loops to provide flow and speed information.

From single loops, speed can be estimated with some advanced techniques, but this kind of speed information should be used with care only. Several recent research efforts are aimed at improving the accuracy of speed estimation and vehicle classification with single loops (Coifman and Kim, 2008). Double loops can determine speeds more easily than single loops and are more easily used for vehicle classification (Heidemann et al., 2008).

Specific studies investigating the accuracy of installed dual loop detectors report unreliable results like underestimation of volumes and false classification while stating that the cause of the inaccuracy could be the hardware, software, or underlying algorithm (Nihan et al., 2002). Traffic volume underestimation was also reported by Briedis and Samuels (2010), who cited pavement condition as the factor producing the highest impact on data quality.

Generally, data from a working and calibrated loop detector are rather accurate and reliable. However, since inductive loops are the typical detectors used for long-term installations, many are broken or biased after long-time usage.

2.2.2.2 Magnetic field detectors

These detectors use the earth's magnetic field to detect vehicles. The metallic mass of a vehicle influences the vertical and horizontal components of the earth's magnetic field locally and this influence can be measured by the sensors. For a description of the method, see Mimbela and Klein (2000). To measure speeds, two sensors within a close distance are needed. In modern detection equipment, both sensors are combined in a single unit. The time series of changes in the magnetic field produced by moving vehicles can also serve as the basis for vehicle classification and patterns for vehicle reidentification.

Since the earth's magnetic field can be distorted by influences such as electric cables, it is necessary to consider these error sources when installing the sensors. A sensor is typically mounted in the middle of a lane on a road surface such that installation and maintenance work can be done without road closure. The systematic disadvantage of magnetic field sensors is that they cannot detect stopped vehicles. Since magnetometers are relatively recent measurement technologies, there is no consensus on their measurement quality.

2.2.2.3 Pressure detectors

A pressure detector can measure the presence of a vehicle at a cross section by measuring the impact of the wheels of the vehicle on the detector. The simplest pressure detectors are thin tubes attached to a road surface. When a vehicle crosses a tube, the air pressure is increased and the pressure can be measured by an electronic device. More advanced pressure detectors use fiber-optic tubes or piezoelectric cables.

Pressure detectors remain the most commonly used sensors for short-term traffic counting and vehicle classification by axle count and spacing. Some types gather data to calculate vehicle gaps, intersection stop delays, stop sign delays, saturation flow rates, spot speeds, and other factors. High truck and bus volumes tend to deteriorate axle count accuracy. Pressure detectors are also prone to breakage from vandalism and wear produced by truck tires (Heidemann et al., 2008).

2.2.2.4 Weigh-in-motion systems and piezoelectric sensors

Weigh-in-motion (WIM) systems are used to capture and record truck axle weights and gross vehicle weights as they pass over sensors. The advantage of state-of the-art WIM systems over older weighting systems is that the vehicles do not have to stop to be weighted. WIM systems use piezoelectric (solid-state quartz) sensors. Forces imposed onto the quartz sensors produce varying electric charge signals that are converted into proportional voltages that can be further calculated to determine wheel loads.

WIM sensors thus show numbers of wheels and axles and weight per wheel, axle, or vehicle. WIM data can be used to calibrate the vehicle type distribution of a simulation model. However, piezoelectric sensors are very sensitive to pavement temperatures and vehicle speeds (Mimbela and Klein, 2000).

2.2.2.5 Passive infrared (PIR) sensors

PIR sensors measure speed and vehicle length as well as volume and lane occupancy. PIR sensors are usually side firing or mounted above the road and detect differences in temperature caused by passing vehicles. The main disadvantage of infrared sensors is sensitivity to weather conditions, including fog, rain, and snow (Mimbela and Klein, 2000).

2.2.2.6 Microwave radar

Microwave radars detect frequency shifts between transmitted and received electromagnetic signals. Two types of radar technologies are used for vehicle detection.

The first type transmits electromagnetic energy at a constant frequency and measures the speeds of vehicles within its field of view using the Doppler principle—the difference in frequency between the transmitted and received signals is proportional to the vehicle speed. The disadvantage of this technology is that stopped vehicles cannot be detected because they do not generate frequency shifts.

The second type transmits a frequency-modulated continuous wave (FMCW) that varies the transmitted frequency continuously over time. It permits detection of stopped vehicles by measuring the range from the detector to the vehicle. Since this detector can sense stopped vehicles, it is sometimes called true-presence microwave radar (Luz and Klein, 2000).

Three radar-based sensors were tested in a field trial described by ATAC (2009) to determine their accuracy in recording traffic volume, vehicle speed, and vehicle classification. In addition, two types of sensor mounting configurations were tested to determine whether they significantly influenced sensor accuracy. The results showed a slight impact of the stability of the mounting configuration on accuracy. The best among the radar sensors tested delivered an overall performance (volume, speed, classification) comparable to a loop detector.

2.2.2.7 Active infrared detectors

Active infrared detectors function similarly to microwave radar detectors. The most common types use laser diodes to transmit energy in the near-infrared spectrum. A portion of the energy is reflected back into the receiver of the detector from a vehicle in its field of view. Laser radars can supply vehicle passage, presence, and speed information. Other types of active infrared detectors use light emitting diodes (LEDs) as signal sources (Luz and Klein, 2000).

2.2.2.8 Ultrasonic (active acoustic) detectors

Ultrasonic detectors transmit pulses of ultrasonic energy toward a roadway. The energy (or a portion) is reflected from the road or vehicle surface into the receiver portion of the device. Based on the changed feedback caused by passing vehicles, the detector notes vehicle passage and presence.

2.2.2.9 Passive acoustic sensors

Vehicular traffic produces audible sounds from a variety of sources within vehicles and from the interactions of vehicle tires with road surfaces. Arrays of acoustic microphones are used to pick up these sounds from a focused area within a lane on a roadway. When a vehicle passes through the detection zone, the signal processing algorithm detects an increase in sound energy and a vehicle presence signal is generated. When the vehicle leaves the detection zone, the sound energy decreases below the detection threshold and the vehicle presence signal is terminated.

When mounted over the center of the roadway and equipped with a fully populated microphone array and adaptive spatial processing to form multiple zones, an acoustic sensor can detect the traffic information for as many as seven lanes. Disadvantages include effects of extreme cold weather on data quality and high purchase and installation costs (Mimbela and Klein, 2000).

2.2.2.10 Cameras and automatic video image processing (loop emulators)

Various available detector systems utilize video cameras to emulate local detectors. Typically, some areas of an image are defined by the user as detection area and the video image processor and subsequent steps of the algorithm derive local traffic parameters like speed, volume, vehicle length, headways, and vehicle classification. In other systems, vehicles are tracked through the entire field of view by identifying and following a path produced by changes in pixel contrast.

Ongoing discussions and continuing investigations surround the quality of video-based vehicle detection. Compared to other techniques, video seems a promising, powerful, and error-prone technology. The most critical factors influencing detection performance are (Grant et al., 1999):

- Lighting conditions: Despite many advances, varying light conditions, shadows, and reflections naturally affect the performance of video-based detectors.

- Camera movements due to unstable installation or strong winds, for example, may lead to reduced detection accuracy.
- Camera position: A camera should ideally be mounted directly above a roadway. Slanted views create problems such as double counting and occlusions.
- Inclement weather leads to reduced visibility.

Yu (2009) reported that false vehicle detections caused by shadows cast by vehicles in adjacent lanes led to problems especially during cloudless weather with strong sunshine. The precision of classification is also susceptible to rainfall, wind gales, and glare from bright light such as direct or reflected sunlight during days and artificial light from car headlights at night.

Video image processing, although considered a "local" data collection technique (and used locally because traffic management centers and statistical departments need specific information), can provide more than local data. The manufacturers claim that their systems can track vehicles and detect lane changes (Versavel, 2007). One advantage of all video-based detectors is that there is—at least at an initial processing step—an image of a traffic situation that may be used for spot-checking traffic conditions.

2.2.3 Nonintrusive detectors: conclusions

Due to their low installation costs, nonintrusive local detectors are used regularly to collect simulation-specific study data sets. For example, one can use cameras triggered by pressure and microwave radar detectors. As the collection periods for these types of data collection are usually short, the data quality and availability must be tested early in the collection process. Indeed, the quality of such data sets is vital for generating good simulation studies. We have seen above that the positioning of many nonintrusive measurement devices exerts a strong impact on measurement quality.

2.3 SECTION DATA

This section starts with a short description of the data characteristics, followed by a more detailed report of the various techniques for collecting these types of data.

2.3.1 Data characteristics

Section data typically reveals (1) the numbers of vehicles traveling from one point in a network to another; (2) travel times between these points, either aggregated in time intervals or for single vehicles; and (3) if all entrances and exits between reidentification devices are detected, the number of vehicles in a section (average density) can be calculated (which is also possible with devices for local data collection).

To measure section data, vehicles must be either reidentified at both points or tracked through the section. Vehicle reidentification basically means tracking the points in time when a vehicle passes several predefined locations in a road network. Depending on possible route alternatives between any two measurement points, an identifiable route corresponds to a pair of measurement points where a vehicle is identified. Methods using vehicle tracking are described in Section 2.4.1.

To reidentify a vehicle, more or less unique feature matrices (scalars, strings, vectors) are extracted from each vehicle as it passes a detector. The feature matrices collected at the various detector stations are compared and matched in a central processing unit. Plausibility checks based on fundamental traffic flow rules and/or common sense can be used to check and increase data reliability.

Using reidentification for OD matrix estimation is somewhat harder than travel time estimation because 100% of the passing vehicles must be detected. If this is not possible, the exact fraction of identified vehicles needs to be known and the OD distribution of the fraction must represent all the vehicles. Vehicle reidentification allows the determination of travel times already at relatively low reidentification rates. These travel times can be compared to simulation outputs. In contrast to local detector data, travel times carry more information about congestion length and severity. Furthermore, travel times are interesting parameters for assessing a transport system from a user perspective.

2.3.2 License plate recognition

Today, the most common technology used for sector-based measurement is license plate reading with cameras (automatic image processing). License plates are truly unique, and all vehicles are equipped with them. The recognition rate today is high enough to provide a reliable volume count and a safe reidentification of vehicles between measurement sites.

2.3.3 Vehicle reidentification based on loop (local) detector signatures

The dynamic change of the electric inductivities of loop detectors during the passage of a vehicle provides a so-called signature, i.e., an inductivity profile of a vehicle over time. Pattern-matching techniques can be used to reidentify vehicles at the next measurement sites. Note that these signatures are not unique. Two vehicles of the same type cannot be distinguished reliably; the sequence of vehicles is used as an additional input for the pattern-matching algorithms. Therefore, this method is better suited for shorter sections of freeways than for longer distances in urban networks.

Several data processing methods have been presented to reidentify vehicles based on inductive loop signatures (Coifman, 1999), and subsequent work supports the vehicle length feature with additional information like travel time and vehicle grouping. The main advantage of this approach is that the required features can be gathered typically via standard interfaces provided by detectors and/or controllers. Other approaches, including those of Ritchie and Sun (1998) and Tabib (2001), use inductive signatures. These signatures provide more information than vehicle lengths but are typically not available via the standard interfaces. Specific controller software is needed, and data can be generated only after close collaboration with detector manufacturers.

Although inductive loops are the most common detector techniques used for vehicle reidentification, other techniques are used as well. Any detector technique that allows extraction of features such as length, height, color, and profile can be used. For example, Remias et al. (2011) describe and evaluate a reidentification system based on magnetometer signature matching. Sun et al. (2004) propose an approach fusing data from inductive loops and video images to enhance reidentification.

Several methods for vehicle reidentification based on vehicle features and signatures are discussed in the literature. Although it is not possible to detail generally expected performance parameters, it is obvious that the following applies:

- The more entering and exiting traffic between two detector stations, the harder it is to reidentify vehicles (making freeway application more feasible than urban applications)
- Some nonintrusive techniques are vulnerable to weather and lighting conditions, making identification especially difficult if conditions vary at the two stations.
- The more homogeneous the extracted features of a vehicle fleet are, the more difficult it is to reliably reidentify a vehicle. For example, if traffic consists of only passenger cars, reidentification based on vehicle length only is difficult if not impossible.

2.3.4 Forthcoming identification technologies

The increase of new communication technologies, protocols, and standards led to the availability of new technologies for vehicle identification (and, hence, reidentification). Some devices are still under investigation; others are commercially available.

2.3.4.1 Bluetooth scanning

Many vehicles provide Bluetooth connections for phone access, and many drivers' mobile phones have activated Bluetooth connections. These Bluetooth signals can be detected outside vehicles, and scanners can read the unique MAC addresses of all Bluetooth devices. Tests show that about 30% of all vehicles are visible, i.e., have their Bluetooth interfaces turned on. The MAC addresses can then be used for there identifications of vehicles. Many recent papers have discussed this relatively new measurement device (Barcelo et al., 2012).

2.3.4.2 Vehicle-to-infrastructure communication based on IEEE 802.11p

Two of the most promising technologies for providing section-based information are the vehicle-to-vehicle (V2V) and vehicle-to-infrastructure (V2I) communications based on wireless LANs, jointly referred to V2X. The technology is not yet deployed for field use, and the first field operational tests are being conducted. It can be expected that in a few years many vehicles will be equipped with V2X devices allowing them to actively report IDs to road-side units that will generate travel time information.

2.3.4.3 Dedicated short-range communication (DSRC)

Dedicated short-range communications involved one- or two-way short- to medium-range wireless communication channels. The European Telecommunications Standards Institute (ETSI) allocated 30 MHz of spectrum in the 5.9 GHz band for intelligent transport systems (ITS) in August 2008. This allows high data rate communications for long distances (up to 1000 meters) with low weather dependence. DSRCs are mainly used for electronic toll collection (ETC; Wikipedia 2011). ETC is a good example of valuable use of technically available traffic data that are not accessible because of administrative issues (e.g., systems are used for other purposes).

2.3.4.4 Commercially available travel time information services

While public authorities and research institutions that may provide traffic data at no cost or via some cooperative arrangement, traffic information can be bought from private companies. Because navigation and mobile phone service providers discovered that their users' data contained a real treasure—TomTom (2011) is one example—the end user can buy travel times per link or route for selectable intervals. In some cases, it may be worthwhile to consider such a service instead of devising and implementing a measurement program. The penetration rate and its homogeneity of distribution within the flow are the keys to reliable travel time measurements with these types of devices.

2.4 TRAJECTORY DATA

Vehicle trajectories are lists of consecutive vehicle positions and recording times. The trajectory data term covers a wide range of data set types, ranging from a single vehicle recording its position every minute (i.e., at 1/60 Hz), to a complete picture of the spatiotemporal distribution of all vehicles in a road section recorded at 25 Hz. Besides accuracy, from the usability perspective, the main differences are (1) trajectories are shown for subject vehicles only and nothing is known about why a driver took a certain action; and (2) trajectories from surrounding vehicles or vehicle groups provide a complete picture, and it is thus possible to relate driver actions to interactions with other vehicles (cars following and lane changes). Trajectory data can be collected by vehicles and by external observations.

2.4.1 Vehicle-based trajectory collection (probe vehicle data)

Vehicle-based trajectories are collected by vehicles equipped with positioning devices, usually GPS. Vehicles equipped with positioning systems are often called probes or probe vehicles. They collect traffic state data as they move through traffic.

Available trajectory data vary based on recording frequency. Dedicated logging devices can be used to record data at high frequencies (1 to 25 Hz). If data from existing fleet management systems (e.g., taxis) are used, recording frequency can be as low as 1/30 Hz or less.

2.4.1.1 Macroscopic traffic parameters

The concept of using probe vehicles to record macroscopic traffic data such as journey time and/or speed has rapidly gained acceptance as a key method for obtaining useful information about highway traffic conditions (Brackstone et al., 2001). The quality of travel time information from probe vehicles depends on the frequency with which they traverse a road link. A large sample of probe vehicles per link per time interval would provide travel time with a high level of confidence. However, the frequency is a function of the number of probe vehicles and distribution of their trips over a network.

To address questions about distribution and frequency of probe vehicles, a detailed understanding of the origin–destination (OD) patterns of probe cars is essential for the development of a travel time prediction model to meet the required levels of accuracy and confidence (Chung et al., 2003). Several studies showed that measurements based on a link or path (Chen and Chien, 2001) can be considered to deliver representative speed estimates with about 1% probe vehicles in heavy traffic.

2.4.1.2 Microscopic traffic parameters

To derive data for microscopic model parameter estimation, a high vehicle position recording frequency is necessary to allow the extraction of information about driving behavior features such as acceleration, deceleration, and lane changing.

2.4.1.3 Quality enhancement

To enhance GPS quality, differential GPS (DGPS) can be employed. DGPS tracks fixed base station receivers at known locations over a certain period. This allows correction of the bias errors of the moving receivers if both the moving receiver and the base station are connected to at least four GPS satellites to centimeter level.

2.4.1.4 Trends

Attempts have been made to use cellular phone signal strengths to estimate vehicle positions based on triangulation. The new Galileo system complements GPS. Additional new sources for obtaining vehicle-based data for calibrating and validating microscopic traffic models are vehicular sensors that are becoming more available with the common use of advanced driver assistance systems (ADAS) such as distance radar, on-board video cameras, and laser scanning devices.

To access the data, dedicated installations are needed in each vehicle. This indicates that vehicular sensor data are mainly available for research activities of manufacturers and/or original equipment manufacturers (OEM). Studies have been carried out with vehicles equipped with such devices. One involved 31 drivers during merging maneuvers (Kondyli and Elefteriadou, 2010). Many naturalistic driving experiments have been conducted since 2010 (Valero-Mora et al., 2013; Chong et al. 2013).

2.4.2 Video-based trajectories

Non-vehicle-based (external) observations for collecting trajectory data are mostly video-based. Image streams are recorded from cameras mounted on relatively high positions above roads. Cameras may be mounted on poles or on airborne vehicles (helicopters, airplanes, airships, drones, satellites, etc.). An airborne system was described by Lenhart et al. (2008).

Video recording can be used to collect time series data of vehicle platoons traveling along a selected road section. The advantages of the method are that data can be collected from large numbers of vehicles and that the drivers are not influenced by the data collection method; therefore, the collected data can be considered unbiased. The data can be used to study both car following and lane changing behavior. In fact, video recording seems to be the only suitable method for studying lane changing behavior.

Hidas and Wagner (2004) describe a configuration with a camera mounted on a pole and a manual reconstruction of the vehicle trajectories based on a frame-by-frame analysis. The trajectories were smoothed using a polynomial function; the accuracy was estimated at 0.5 m.

As part of the Next Generation Simulation (NGSIM) program (NGSIM, 2011), trajectory data were collected and made available to the public. Further information on the collection and processing of image sequences can be found in Section 2.2.2.10 and in more detail in Hoogendoorn et al. (2003) and subsequent work. Additional valuable sources for vehicle trajectories are data collections from Europe and research projects such as DRIVE C2X (2011).

Video-based trajectories present three main drawbacks. First, if only one camera is used, the observation covers minimal time and space and the consistency of driver following and lane changing behaviors cannot be evaluated. From a practical view, it is often difficult to find suitable high vantage points for positioning cameras. The second drawback is that data postprocessing is very difficult and resource-intensive unless automated image processing methods can be used and they are labor-intensive. The third drawback is the noise in the resulting trajectories. Indeed, (x,y) position measurements result from a complicated process of identifying moving objects on a set of successive images and the result may exhibit lower than expected precision. In addition, errors in position measurement become larger errors in speed (and even in acceleration). Careful filtering is needed, and the relevant techniques are presented in Section 3.2.

2.4.3 Mobile phone data

Increasing numbers of drivers have mobile phones in their vehicles to hold connections to base stations so the drivers are reachable to callers. It therefore seems attractive to use mobile phones as tracking devices for travelers. The localization of the phones, however, is not as exact as desired, especially when phones are not transmitting calls or data. The exactness of the localization depends on the technical equipment used by the telephone provider.

Several research projects investigated the use of mobile phone data (also called floating phone data, or FPD) for traffic detection on small and large scales to derive congestion and origin–destination information. At present, the availability of floating phone data is rare and the quality is difficult to judge. FPD may be useful for studying OD relations, route choice, and travel times. Overviews of the technology and its use can be found in Schlaich (2010) and Borzacchiello (2009).

2.5 STUDIES OF DRIVER AND TRAVELER BEHAVIOR

Studies of the behaviors of drivers and travelers are used to gain insight into specific aspects of the driving and traveling processes. On a driving level, the operational or tactical behaviors, including car following, lane selection, and lane changing are observed and associated with additional data that allow explanations of certain reactions (shifting, steering, and drivers' eye movements recorded by cameras).

Driving behavior data can be collected in driving simulator studies or by field trials, i.e., real-world observations using specially instrumented vehicles. A use case may examine the reaction of a driver to a newly developed in-vehicle information device. Detailed modeling of driver behavior considers errors in driver observation, decision making, and actions taken. Drivers' limited resources of attention in relation to the complexity of the traffic environment and the vehicle equipment should also be considered. A driver's attention to driving tasks and secondary visual tasks (such as monitoring vehicle equipment) can be measured by recording eye movements. However, the attention resources used for a cognitive secondary tasks (thinking of something other than driving) cannot be measured; only the consequences can be reviewed.

By using an instrumented vehicle along with eye movement recording, it is possible to calibrate driver attention distributions at least for secondary visual tasks.

The use of driving simulators offers more possibilities for devising critical conflict situations than are possible with instrumented vehicles. The literature provides many studies related to driver safety and attention. Studies facilitate experimental control and data collection, increase efficiency and safety, and reducing costs (Bella, 2005, 2008). They are useful for testing roadway delineations (Molino et al., 2005) and for testing roadway treatments that involve raising pavements such as center line rumble strips (Auberlet, 2010).

2.5.1 Driving simulators

Although flight simulators have been applied since the early 1900s, driving simulators appeared in primitive forms only in the 1970s. Distinctions among different simulators are based on levels (Kaptein et al, 1996):

- Low-level simulators typically consist of PCs or graphics workstations, monitors, and simple caps with controls.
- Mid-level devices include advanced imaging techniques, large projection screens, realistic caps, and perhaps simple motion bases.
- High-level systems typically provide almost 360-degree fields of view and extensive moving bases.

2.5.2 Advantages and limitations of driving simulators

The applications for driving simulators include human factor research, medical research, driver education, training, assessment, and evaluation of vehicle design. Among the many reasons or advantages of using driving simulators in research on traffic behavior (Kaptein et al, 1996) are:

- Enabling the investigation of effects of nonexistent road elements
- Economy; new road designs are too expensive to build just to test effects on driver behavior
- Ability to investigate dangerous situations without injuries
- Exposure to events that may occur rarely in practice
- Avoidance of liability encountered with road tests
- Optimal experimental control

In spite of the advantages of driving simulators, they have one main disadvantage: the driving task in a simulator cannot be completely realistic. Therefore, any use of driving simulators should be preceded by determining whether the simulator is sufficiently valid for the task to be investigated. Because people rarely need all available information to perform a task, it is generally not necessary that the available information about a simulator be identical to what would be available in a real vehicle (Flexman and Stark, 1987). In some cases, a deliberate deviation from reality may produce more realistic task performance.

Recent progress in electronic and computer technology has made possible the development of relatively low-cost, laboratory-based driving simulators. Driving simulators provide a safe and economical way to test skills during adverse road scenarios (Rizzo et al., 2001). Simulators are widely used internationally to study driver behaviors (Blana, 1996). Many studies concluded that driving simulators can provide accurate observations of driver behaviors and functions (Alicandri, 1994; Fraser et al., 1994; Van Winsum, 1996; Desmond and Matthews, 1997; Ellingrod et al., 1997; Van Winsum and Brouwer, 1997).

2.5.3 Simulator validity

Leung and Starmer (2005) distinguish absolute and relative validity and internal and external validity. These and other types of validities are explained below.

Absolute validity — The simulator has absolute validity with regard to a research question if the absolute size of the effect is comparable to the effect in reality.

Relative validity — The simulator has relative validity with regard to a research question if the direction or relative size of the effect of the measure is the same as in reality.

Internal validity — The recognition of a possible apparent relationship between a manipulation and an obtained effect. A research method has internal validity if an obtained effect has no alternative explanations. Internal validity may be lost if driver behavior is specifically affected by the limitations of a driving simulator, such as:

- Limited resolution of a computer-generated image
- Delay until vehicle position and images are updated
- Restricted horizontal field of view
- Little or no proprioceptive information

External validity — The extent that the results obtained from a specific set of subjects in a specific environment during a specific period can be generalized to other persons, environments, and time periods. Limitations in external validity may, for example, be caused by (1) careless choice of test environment and (2) subject selection (driver experience, motivation, mental and physical condition, and other factors).

Face validity — How realistic an experimental environment like a driving simulator appears to the subjects.

Statistical conclusion validity — The extent to which results are statistically reliable and the statistical tests and estimations correctly applied.

Construct validity — A type of external validity that determines whether a method of measurement actually measures the construct intended to be measured.

The basis of a valid research project—with or without the use of a driving simulator—is a carefully designed experiment (internal validity). Tornos (1998) observed that, for a simulator to be a useful research tool, relative validity is necessary, but absolute validity is not essential. This is because research questions usually deal with matters relating to the effects of independent variables rather than seeking absolute numerical measurements of driver behaviors.

Kaptein et al. (1996) conducted a survey of validation studies on a fixed base driving simulator manufactured by TNO (Soesterberg, Netherlands). In these validation studies, the behaviors of drivers in simulators were compared to real-life driving behaviors. From research conducted by Blaauw (1984), it followed that drivers in the simulator drove faster and showed more variations in lateral positions than drivers in real life. Another study conducted with this fixed base driving simulator (Kaptein et al., 1996) showed that the available information is crucial in analyzing the validity of driving simulators. When drivers had more visual information, they performed better on a braking test. This study established the relative validity of driving simulators.

Validation studies were also performed using a Daimler-Benz simulator in Berlin. An experiment conducted by Riemersma et al. (1990) indicated drivers in the simulator adjusted their speed more in reaction to speed-reducing measures than in real life. The study also concluded that the driving simulator possessed relative validity.

Some studies utilized the fixed base Leeds Advanced Driving Simulator (LADS). Carsten et al. (1997) investigated the speeds and lateral positions of vehicles at 21 locations. Mean speed in the driving simulator did not differ significantly from real-life behavior. For lateral positioning, a significant difference could be observed between reality and driving simulator. However, qualitatively the same behavioral tendencies could be observed.

Recent research of Yan et al. (2008) also showed that driving simulators can be assumed to have relative validity. Their experiment compared differences between behavior in the driving simulator and at real-life crossroads. Reimer et al. (2006) used self-reports to compare behaviors in a simulator to real-life driving behavior. Significant relationships were established with regard to accidents, speeding, speed, overtaking, and behavior at traffic signs. These studies indicate that driving simulators generally possess task-specific relative validity.

In the real world, driving is a goal-orientated task. This implies that driver behavior is the result of a balance of perceived benefits (i.e., arriving at work on time) and perceived costs (i.e., driving risks). These motivations are critical to engendering realistic driver performance in simulators in terms of tactical and knowledge-based behaviors. For example, the degree of motivation provided in a driving simulator may affect the headway and speed choices of drivers and their route choice decisions. Thus, realistic motivations in driving simulators are critical components of the validity of the observed results and caution is required if we wish to extrapolate simulator results from a laboratory to a real-life driving environment (Greenberg, 2004; Neimer, 2009).

2.5.4 Simulator sickness

Both simulators and virtual environments (VEs) can cause certain types of sicknesses and other physical problems. Examples are visuomotor dysfunctions (eyestrain, blurred vision, focusing difficulty), mental disorientation (concentration difficulty, confusion, apathy), and nausea and vomiting. Other symptoms include drowsiness, fatigue, and headache. The Motion Sickness Susceptibility Questionnaire that has been used successfully to assess the susceptibility to motion sickness (Reason and Brand, 1975) could be adapted to screen susceptible elderly drivers prior to simulator testing.

The two requirements for simulator sickness are a functioning vestibular system (the set of inner ear canals, tubes, and other structures that sense orientation and acceleration) and a sense of motion. There is no definitive explanation for simulator motion sickness, but one idea is that it arises from a mismatch between visual motion and physical cues as perceived by the vestibular system. This can happen when there are no physical motion cues (no motion platform is used) or the physical and visual cues are not synchronized. In VE systems, simulator sickness occurs both in motion-based (game pods) and physically static devices. One hypothesis for the onset of the sickness is that these inconsistent perceptions are similar to what occurs when poison is ingested; human evolution allows us to vomit to eliminate the poison. Studies show that other factors are more likely to cause sickness:

- Bright images rather than dim (nighttime) images
- Wide fields of view
- CRTs rather than dome projection systems
- Nauscogenic motion (at 0.2 Hz)

2.5.5 Use of simulators to model driver behaviors

Driving simulators are often used in the development traffic models, mainly at the microscopic level. An example of the use of the driving simulator studies for traffic simulations was the European Community's STARDUST project (2001–2004). Figure 2.1 illustrates this use. One the goals of STARDUST was to assess the extent to which advanced driver assistance systems (ADAS) may contribute to a sustainable urban development based on traffic capacity. For this type of application, data collection is problematic, and simulators are often used as data providers for microscopic simulation parameter calibrations.

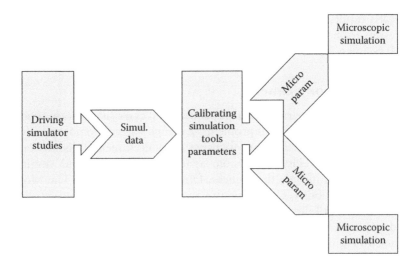

Figure 2.1 Example of use of driving simulator studies for traffic simulation.

Farah et al. (2009b) used a driving simulator to collect detailed trajectory data of passing maneuvers on two-lane rural highways and then developed a passing gap acceptance model that accounts for the impacts of road geometry, traffic conditions, and driver characteristics. Farah and Toledo (2010, 2011) investigated various definitions of passing gaps and used the definitions to develop three passing gap acceptance models that were calibrated based on passing maneuver data collected by a driving simulator. The generic structures of these models utilize drivers' desires to pass and their gap acceptance decisions. Impacts of traffic characteristics, road geometry, and driver characteristics were included in these models.

Farah et al. (2009a) also developed a model that explains the minimum time to collision. The model formulation is based on the analysis of drivers' passing decisions on two-lane rural highways using an interactive driving simulator. The simulator allows collection of vehicle speed and position data for road and traffic scenarios. In addition to simulator activities, participants responded to a questionnaire that sought information about their sociodemographic characteristics. The composed data set was analyzed and processed to develop a model that predicts the risks associated with passing behaviors.

Tobit regression models were found to be more suitable in comparison to ordinary least squares models and hazard-based duration models. The tested explanatory variables represented road geometry, traffic conditions, and driver characteristics. While the traffic-related variables produced the most important effect on the chosen risk measure, factors related to geometric design and the driver characteristics also made significant contributions.

Driving simulators were used recently to study the influences of adverse conditions on adaptation effects in longitudinal driving behavior and the determinants of these adaptation effects. Examples of adverse conditions are incidents, adverse weather conditions, and emergency situations.

Hoogendoorn et al. (2010) demonstrated that perception of an incident (i.e., car accident) in the advanced driving simulator at Delft University of Technology led to substantial adaptation effects in longitudinal driving behavior (reduction in speed, acceleration, deceleration, and an increase in distance headway). Furthermore, perception of the incident led to an increase in mental workloads of drivers, as indicated by the physiological indicators of effort expenditure. Parameters of the intelligent driver model (Treiber et al., 2000) and the Helly model (Helly, 1959) were estimated using a new calibration approach for joint estimation (Hoogendoorn and Hoogendoorn, 2010b). They showed that substantial parameter value changes and a reduction in model performance could be observed in the vicinity of an incident.

Comparable results were found with regard to adverse weather conditions (Hoogendoorn et al., 2011). Fog simulated in the driving simulator exerted considerable influence on longitudinal driving behavior and led to a significant increase in mental workload.

Finally, Hoogendoorn et al. (2011) utilized a driving simulator study to determine the extent to which emergency situations led to adaptation effects in longitudinal driving behavior, changes in mental workload, and changes in the positions of action points in psycho-spacing models. Again significant adaptation effects in driving behavior, significant increases in mental workload, and substantial changes in the positions of action points in psycho-spacing models were established.

Drivers' responses to other vehicles and traffic signals are predictable. However, when real-time traffic information and route guidance are involved, responses are more difficult to estimate. Therefore, more data are needed about the driver and traveler responses to specific types of traffic information and guidance.

2.6 STATED AND REVEALED PREFERENCES

Stated preferences and revealed preferences can be used to determine and explain travel behavior on a more strategic level. Stated preferences usually describe scenarios and test persons are asked how they would react in certain situations (e.g., how they would change a mode choice if bus travel time were reduced due to a new express line and the time savings were offset with a fare increase).

In revealed preferences, actual empirical travel behavior is recorded, usually as protocols or logbooks. Smart phones with GPS and acceleration sensors can be used to collect travel behavior data from volunteers. While the GPS provides a route from origin to destination, an accelerometer can be used to identify the mode of transport (walking, driving, public transport, etc.). Volunteers' background variables (age, gender, etc.) are also recorded. In both types of experiments, the investigator must correctly define the sample of interviewed users. Their representativeness of a complete population impacted by a measure (for example, reduction of bus travel time cited above) is key for a correct determination of global behavior.

2.7 METADATA

Certain types of data are traffic-relevant but do not constitute actual traffic data. They are traffic-relevant because they may play roles in the processes of calibrating and validating simulation models. They can be distinguished in simulation input and output data.

Input data need to be considered when calibrating or validating simulation models as they impact traffic and travel behavior. When comparing a simulation model to real-world data, the context must be considered, for example, by classifying traffic data with respect to traffic-relevant data based on various parameters such as precipitation intensity. Weather data and road surface conditions are known to produce large impacts on driving behavior (El Faouzi et al., 2010) and include:

- Precipitation intensity
- Precipitation type
- Visibility
- Water film thickness
- Road surface condition

These data are typically collected locally along major roads and used for traffic control and/or winter maintenance operations. Several research projects collected road weather and road surface data using vehicle sensors. The COST TU0702 study produced a detailed report on the influence of adverse conditions on traffic and safety (El Faouzi et al., 2010; COST TU0702, 2011).

General weather conditions affect travel behavior (mainly activity chains, mode choices, route choices, and departure times). The general weather situations observed and recorded by the national meteorological services can be used, for example, when conducting a survey on mode choice and how rain affects choices. Certain types of data that correspond to output data may be estimated by a simulation and hence need to be compared to real-world measurements. Of most importance are emission and pollution data. Emissions data are detected mainly by instrumented vehicles. Pollution emissions are measured by roadside stations.

Site-specific static data related to the traffic environment also need to be collected for detailed traffic simulation and calibration operations. The static structure of a traffic environment affects the capacity of a road section or intersection. The static data may include road type and width, lane dimensions, and ramp lengths. Intersection data should detail type (signalized, nonsignalized, roundabout), geometry, pocket lanes (lengths), pedestrian crossings, conflict points within the intersection, and yielding rules. In signalized junctions, the arrangement of traffic signals and detectors along with signal timing and phasing must be known before simulations, calibration, and validation can be completed. The static data may also involve sight distances and objects such as buildings, signposts, vegetation, and other objects that limit visibility.

The operation of modeling traffic control and management systems incorporated into a simulation model should also be validated unless hardware in-the-loop simulation is used. Data covering signal control systems should be collected because these systems exert considerable impacts on the capacities of junctions, corridors, and even networks. Since the operations of various signal control systems can differ widely, data about the control principles and control parameters should be compiled, then the operation of the signal control within the simulation model should be verified, calibrated, and validated.

2.8 CONCLUSIONS

2.8.1 Useful data

As expected, there is no general answer about data to be used in modeling. The answers depend on the type of calibration and validation tasks and ultimately data availability.

In traffic engineering projects where models must be calibrated to replicate macroscopic traffic features and budgets are usually tight, the first principle is to use what is available. Relevant data may already be sufficient for freeways with dense networks of local detectors. It is important to note that demand data (depending on the system volume at the network boundaries or a valid OD matrix) and data reflecting service levels (travel times, speeds) must be available.

Travel time and speed information can be collected by one or more vehicles equipped with low-cost nomadic devices that include GPS (free software for postprocessing GPS tracks is available through the Internet) or by a pair of vehicle identification devices (ANPR cameras). The vehicle-mounted devices are probably better if only a few observations are needed. ANPR equipment involves higher initial costs but is more effective if multiple or lengthy observation periods are required.

To investigate driving behavior (car following, lane changing), it may be worthwhile to consider using an existing data set if it is available and fits the need. Information platforms listing available data sources include the MULTITUDE website (with a continuously updated overview), the NGSIM program site, and the large-scale field operational test (FOT) projects such as DRIVE C2X.

2.8.2 Expected data quality

Data quality should be considered when drawing conclusions based on the data. Therefore, it is desirable to have a quality or accuracy estimate for the various data types and detection techniques. However, this is nearly impossible to achieve because the quality of a device depends on various factors related to hardware, software, system configuration, correctness of installation, deterioration, and maintenance intervals.

Although standards and guidelines implemented in many countries define requirements for data quality, especially related to local traffic data, we cannot say that data gathered from a device in the field will meet the defined quality standard. It is far too costly to evaluate and maintain the quality of every installed detector.

Traffic data should always be investigated from a traffic engineering view. Does the fundamental diagram appear correct? Is the vehicle conservation law fulfilled? Does the time series look plausible? A lot of information about data quality is typically available for well known datasets such as the NGSIM trajectories.

2.8.3 Suggestions for further actions

The list below covers suggestions for further research and analysis of data collection techniques. These were explored by WG2 in the MULTITUDE project. Furthermore, the state-of-the-art data collection techniques showed the need for some practical actions.

2.8.3.1 Macroscopic data

Local detector data reference studies — There are a lot of studies and comparisons available, with a recent focus on nonintrusive techniques. Although (or because) a lot of information is available, it is not easy to obtain a clear picture of the best circumstances and techniques for a specific purpose. It would be helpful to have a sort of "metareport" incorporating the findings of the relevant studies.

Local detector data — Although up-to-date communication and data storage infrastructures easily allow collection of individual vehicle data, in most cases data are aggregated at a local controller, making individual vehicle data hard to obtain. Individual vehicle data are far more valuable for microscopic model calibration and validation for other purposes. It would be very desirable to consider the need for individual vehicle information in future standardization and system layout procedures.

Electronic toll collection data — Travel time data from these systems are potentially very useful as a validation basis. It may be worthwhile to explore how to overcome administrative and organization hurdles in order to make these data available.

2.8.3.2 Microscopic data

Data sets for microscopic model calibration — Such data sets are extremely costly to compile and invaluable for the derivation and validation of simulation models. There exist several apparently highly useful and elaborate data sets. It would be desirable to strengthen efforts to make them more easily understandable and available, possibly in the scope of research networks, to make model development more efficient.

Advanced driver assistance systems (ADAS) data and other data collected through EU-funded projects — Data from instrumented vehicles are collected for many large-scale EU projects involving vehicle manufacturers and original equipment manufacturers. It would be sensible to explore mechanisms to make these data available for other EU research activities.

2.8.4 Overview table

This chapter distinguished six data types (local detector, section, trajectory [vehicle- and video-based], behavioral, stated, and revealed preference data). Table 2.1 provides an overview of the six data types and lists the traffic parameters that can be estimated using them.

Table 2.1 Overview of six data types and their characteristics

Data type	Local detector data	Section data (vehicle reidentification)	Vehicle-based trajectory data	Video-based trajectory data	Behavior and driving simulation	Stated and revealed preferences
Traffic variables	Aggregated: speed, type, flow, speed variance, mean headway, mean length, occupancy Individual vehicle: speed, class, length, headway, signature	Travel time, path flow (typically subset of all vehicles)	Position, time stamp, speed, acceleration, lane (typically for subset)	Position, time stamp, speed, acceleration, lane (typically for all vehicles in defined segment)	Position, time stamp, speed, acceleration, lane (typically for single driver or vehicle)	Travel patterns (OD demands, estimations of disaggregate demands, route choices)
Data collection techniques	Intrusive: loop detectors Nonintrusive: magnetic field sensors, radar, acoustic, active and passive infrared, ultrasonic, pressure detectors, cameras	Automatic number plate recognition, Bluetooth, detector signature, DSRC, IEEE 802.11p, commercial services	GPS, DGPS, mobile phones	Video camera and image processor	Virtual vehicle trajectories	Interviews
Calibration and validation uses cases	Macroscopic: traffic demand at boundaries, local traffic state, vehicle types (fleet mix) Individual vehicles: derivation of free speed and headway distributions	Macroscopic: overall validation of specific simulation model	Macroscopic: links path travel times, numbers of stops, queue lengths Microscopic: car following, lane changing	Macroscopic: links travel time, numbers of stops, queue lengths Microscopic: car following, lane changing	Microscopic: car following, lane changing; includes additional data about driver behavior based on instruments and/or interviews	Macroscopic: travel patterns (OD demand, estimations of disaggregate demands, route choices)

Quality issues	Widely studied, many reports available, automatic quality monitoring methods available (sometimes in data logs, quality can be checked with traffic engineering expertise)	Filtering and outlier detection algorithms needed and available; expected quality highest with unique feature vector (e.g., license plate)	Variable; dedicated measurement campaigns typically lead to higher quality than using, e.g., data from a taxi dispatch system	Existing data sets were checked extensively and have high quality; if a new data set is required, manual postprocessing is still necessary	Critical question is whether data are representative of genuine behaviors	Critical question is whether data are representative of genuine behaviors
Data availability	High availability: For specific study areas typically available via local transport authorities or traffic management centers	Deploys far fewer systems than local detection systems; if available, obtain from local transport authority or traffic management center	Via MULTITUDE survey of available data sets or dedicated data collection	NGSIM database or possibly research institutes that maintain their own databases; data collection is labor intensive	Data should be available at various institutes maintaining driving simulators; studies are expensive to plan and perform and should involve specialists such as psychologists	Expensive planning and implementation

(continued)

Table 2.1 Overview of six data types and their characteristics (Continued)

Data type	Local detector data	Section data (vehicle reidentification)	Vehicle-based trajectory data	Video-based trajectory data	Behavior and driving simulation	Stated and revealed preferences
Specific advantages	Many reports available, easy to interpret and process	Not many measurements needed to provide good understanding of system functions; easy to interpret and process	Can reveal microscopic parameters to investigate driving behavior and interactions with other drivers if additional onboard sensors or probe vehicles are available	Gives a complete picture of driving behavior and interactions with other drivers; the real "ground truth"	Controlled environment; same experiment can be repeated; experimental vehicle equipment (ADAS, information systems) can be investigated	Provides data that cannot be gathered by other means
Potential disadvantages	Cover only local and typically aggregated traffic parameters	Typically macroscopic, no insight into local effects	Covers only small subset of all drivers, labor-intensive processing	Covers short sections; no observations of drivers over long periods; very labor intensive	Doubt about whether this "virtual trip" represents real driving behavior	Planning and implementation are expensive; stated preferences may not reflect real behaviors

Chapter 3

Data processing and enhancement techniques

Serge Hoogendoorn, Constantinos Antoniou, Thomas Schreiter, Yufei Yuan, Jaume Barcelo, Christine Buisson, Vincenzo Punzo, and Hans van Lint

CONTENTS

This chapter describes techniques to filter, aggregate, and correct data. Chapter 4 shows the essential role of data in the calibration and the validation of simulation parameters and discusses the impacts of data errors on calibration results. Therefore, before introducing calibration and validation, it is worthwhile to present the main techniques to process and enhance data.

3.1 INTRODUCTION

The application of models, off-line (assessment studies) or on-line (short-term predictions), requires high-quality, processed data that provide models with the required inputs. In many cases, raw data generated by different kinds of sensors cannot be used directly for this purpose, and additional data processing is needed. This chapter deals with this step, that is, with data processing and enhancement techniques for preparing the input data for off-line and on-line model applications.

3.1.1 Data processing applications for simulation studies

For off-line applications, the main purposes of this step are summarized as follows:

- Making data applicable for model calibration and validation, e.g., by means of data processing or data fusion, that is, multiple data sources are combined into a consistent picture of a traffic situation.
- Deriving the key features and patterns necessary for model verification (face validity) and calibration (construct validity).
- Estimating the key input variables (off-line OD tables that may be static or dynamic, depending on the model requirements) from available raw data, e.g., traffic counts.

The main purposes of the data processing and enhancement steps for on-line applications are:

- Turning data into a consistent estimate of the current state of a system (state estimation) that is useful for prediction purposes; in many cases, the short-term prediction methods require accurate estimations of a current traffic situation.
- Deriving key patterns in the data needed for decision making and inferring from these patterns the decisions that need to be made.
- Determining key inputs for prediction purposes (on-line dynamic OD estimates).

In sum, for off-line applications, the key input data are the OD tables and boundary conditions, while information about the traffic state is used for model calibration and validation purposes. Note that we distinguish calibration of model input (the focus in this chapter) and calibration of model parameters (the focus in Chapters 4 and 5).

3.1.2 Purpose-based taxonomy on the purpose of state estimation and data fusion techniques

We will refer to the process of inferencing from raw data as (state) estimation. The estimate can be made using a single source of data (e.g., loop counts) or by using multiple data sources jointly (e.g., loop counts and floating data). In the latter case, the process is usually called data fusion.

Before discussing various techniques, let us look at the different types of inferences that can be distinguished (Varshney, 1997). We will consider three levels based primarily on state estimation as outlined in Table 3.1.

Table 3.1 Inference levels for state estimation and data fusion

Level	Purpose	Typical techniques
1	Processing raw data	State estimation methods: digital filters, least squares estimation, expectation–maximization methods, Kalman filters, particle filters
2	Deriving features and patterns	Classification and inference methods: statistical pattern recognition, Dempster-Shafer, Bayesian methods, neural networks, correlation measures, fuzzy set theory
3	Making decisions and detecting events	Decision support systems (DSSs) and expert systems (Bayesian belief networks, fuzzy methods, artificial intelligence)

The table shows common techniques used to perform state estimation and data fusion at each level.

The first-level estimation and data fusion technique is targeted at raw data processing and estimates the key microscopic or macroscopic traffic flow variables. The quality of estimation is determined by both the accuracy and number of measurements and by the state estimation or data fusion techniques. The methods used at Level 1 may be combinations of digital filters, model-based (e.g., Kalman) filters, simulation techniques (e.g., particle filters). The results from this level general and so forthly constitute the basis for the higher-level estimations. In this chapter, the approaches discussed include Kalman filtering, particle filters, and the adaptive smoothing method (ASM) of Treiber and Helbing (2002).

Level 2 is aimed at deriving features and patterns from the traffic state estimates. For instance, when flows and speeds have been estimated by means of Level 1 state estimation techniques, features such as the shock wave speeds, parameters of the fundamental diagram, minimum distance headways, and others can be inferred from the state estimates. For this purpose, several statistics and inference methods can be applied. The common ones are Bayesian methods, neural networks, regression models, fuzzy logic, maximum likelihood, and others. Training and learning processes may be involved but are not necessarily required. This chapter will present several examples of inferencing. All of them revolve around determining the parameters of traffic flow from the data (shock wave speeds, capacity estimation, OD estimations).

Level 3 may be regarded as the decision level. State estimation and data fusion from the first two levels provide evidence concerning specific events within the observed system. This evidence may trigger a certain decision or initiate a chain of actions. On this level, more human activities are involved to generate many subjective evaluations and assessments. The common approaches on this level are decision support systems (DSSs) and expert systems.

At all levels, the data used and the results of data fusion may change from exact figures to language specification, and the difficulty in data fusion increases. In addition, more human effects are involved at higher levels and lead to more uncertainties.

3.1.3 Structure of chapter

In the remaining sections, we will provide an overview of techniques pertaining to these different levels. The focus will be on Levels 1 and 2; examples of Level 3 data fusion are less relevant for the purpose of this book. Table 3.2 shows an overview of the topics discussed in this chapter.

Table 3.2 Descriptions of data topics and purposes

Topic	Level	Purpose
Microscopic data enhancement	1	Filtering microscopic data (e.g., vehicle trajectories)
Macroscopic traffic state estimation	1	Deriving traffic states (macroscopic) from single or multiple data sources
Traffic flow feature extraction	2	Determining key flow characteristics (e.g., shock wave speeds and capacities) from state estimates
OD estimation	2	Estimating static or dynamic OD tables

3.2 MICROSCOPIC DATA ENHANCEMENT TECHNIQUES

As will be shown in Chapters 4 and 5, parameter estimations of microscopic traffic flow models may rely on time series of speeds or counts at detectors, respectively, for simulation- and model-based approaches or on trajectories of vehicles. In this section, data processing and enhancement techniques for trajectory data used in parameter estimation will be discussed. Techniques for detector data are presented in the next section.

As noted in Section 2.4, obtainable measurements of a vehicle trajectory consist of discrete observations of its position, equally spaced in time, that is, the series of vehicle coordinates in the two- or three-dimensional space. Such measurements are generally derived from video images or directly gathered by GPS devices. After raw positions of vehicles have been determined, additional processing is needed to achieve the accuracy required for the specific application, particularly for deriving variables like vehicles speeds and accelerations.

Note, however, that GPS data may come with an additional independent speed measurement; this is typically the case for so-called RTK-GPS, and it is measured by using the Doppler shift between the device and the satellites. Therefore, speed is not derived from the difference in distance traveled between two subsequent raw positions, which is a strong advantage that often minimizes the error in a GPS-based trajectory estimation.

Section 2.4 presents an overview of trajectory data collection and processing studies and discusses the processing of video data to determine vehicle positions as well. Whatever the technology applied, after raw positions of vehicles have been obtained, they must be processed (enhanced or smoothed) to be of any use. In addition, to derive any other variable of interest, position data accuracy has to be enhanced, reducing the impacts of measurement errors. Indeed, depending on the aim of a study, the potential impacts of measurement errors on results may be significant.

For example, Ossen and Hoogendoorn (2008a) verified the effects on car-following model calibration of adding to vehicle displacements the errors from different distributions accounting for various combinations of random and systematic components. They basically pointed out that measurement errors yield considerable bias in calibration results.

3.2.1 Trajectory data enhancement

The first step in trajectory estimation is to reconstruct from sequential points the trajectory of the observed vehicle, $\hat{s}_n(t)$, that is, the function that gives the space traveled along its path by vehicle n until instant t. From the trajectory function, every kinematic characteristic such as speed and acceleration of the vehicle motion can be derived.

However, when gathering vehicle positions, the observed points happen to be dispersed in the neighborhoods of the actual points and the path drawn on the observed positions inevitably fluctuates around the true path followed by the vehicle. Such measuring errors, when considering two consecutive vehicle positions, propagate into the value of the space traveled between these two points, that is, into the estimated vehicle trajectory function, $\hat{s}_n(t)$ (Punzo et al., 2011).

First, we may observe that the space traveled by a vehicle in two consecutive measurement intervals can differ greatly, up to unphysical values of this difference. This effect becomes clearly visible as a noise in $\hat{s}_n(t)$ that is even greater in the speed and acceleration functions since the differentiation process magnifies the noise despite the numerical scheme adopted. This is the random component of the error in the space traveled.

Second, the estimated cumulative space traveled by vehicle n until instant t, i.e., $\hat{s}_n(t)$, is generally greater than the true one, resulting in a structural error (called bias).

It is worth noting that, unlike the noise, the bias of $\hat{s}_n(t)$ is not self-evident if we look at the trajectory of a single vehicle, but it may become visible if we concurrently deal with the trajectories of a pair of following vehicles. This consideration led to the definition of the so-called platoon consistency (Punzo et al., 2005). This criterion verifies whether trajectories of following vehicles are consistent with regard to space traveled (Punzo et al., 2011).

The idea behind platoon consistency is that when the point coordinates are projected along a local reference system (with one axis aligned to the road) we obtain the longitudinal component of the total space traveled. Since this component is not affected by the bias described above, in all the cases where it is acceptable to confuse the actual vehicle path with its projection on the lane alignment as in the case of pure car-following studies, the projection of the coordinates on the lane alignment is the basic way to remove the bias in space traveled.

Regarding the noise in the positions reported above, the application of filtering and smoothing techniques is compulsory because the noise magnifies in the differentiation to speed and acceleration. In addition, the differentiation process must preserve the basic equations of motion, i.e., estimated speeds and accelerations after integration must return the measured traveled space. Toledo et al. (2007) called this condition internal consistency. Note that the definitions of both internal and platoon consistency provide quantitative tolls to measure the accuracy of trajectory data.

Thiemann et al. (2008) pointed out that the order of differentiations and filtering operations significantly impacted results and noted three possible task orders: (1) smooth positions and then differentiate to velocities and accelerations, (2) differentiate to velocities and accelerations and then smooth all three variables, or (3) smooth positions, differentiate to velocities, smooth velocities, differentiate to accelerations, and smooth accelerations.

3.2.2 Overview of applied filtering techniques

Table 3.3 reports and classifies the studies addressing the problem of filtering trajectory data. Columns are arranged in increasing order of complexity—from averaging to Kalman smoothing—while rows point at the variable to which each technique was applied.

Table 3.3 Techniques cited in literature

Variable	Averaging	Smoothing	Kalman filtering	Kalman smoothing
Coordinates (x,t)	Hamdar and Mahmassani (2008)		Ervin et al. (1991)	
Trajectory $s(t)$	Ossen and Hoogendoorn (2008a); Thiemann et al. (2008)	Punzo et al. (2005); Toledo et al. (2007)		Ma and Andreasson (2006)
Intervehicle spacing $\Delta s(t)$			Punzo et al. (2005)	Ma and Andreasson (2006)
Speed $V(t)$	Hamdar and Mahmassani (2008); Thiemann et al. (2008)	Brockfeld et al. (2004); Punzo et al. (2005); Lu and Skabardonis (2007); Xin et al. (2008)	Punzo et al. (2005)	Ma and Andreasson (2006)
Acceleration $A(t)$	Thiemann et al. (2008)	Xin et al. (2008)		

Source: Punzo, V., Borzacchiello, M.T., and Ciuffo, B. 2011. On the assessment of vehicle trajectory data accuracy and application to next generation SIMulation (NGSIM) program data. Transportation Research Part C 19, 1243–1262.

The most basic technique among those cited in the table is the simple moving average. Ossen and Hoogendoorn (2008a) apply it to synthetic noisy data. Hamdar and Mahmassani (2008) apply a monodimensional Gaussian kernel moving average to eliminate discontinuities in lateral x-coordinates and speeds from NGSIM data. They report minor discontinuities after smoothing.

Thiemann et al. (2008) use a symmetric exponential moving average filter to smooth NGSIM data and propose to address the problem of noise magnification after differentiation, first by differentiating positions to speeds and accelerations and then by smoothing all three variables. It is worth noting that the moving average generally needs a sensitivity analysis to find windows and weights that provide the best trade-off between the damping of the noise and the losses of high-frequency parts of original data.

Among those who apply smoothing techniques, instead, Brockfeld et al. (2004) calculate the accelerations from speeds from DGPS data by means of a Savitzky-Golay smoothing filter, with a second-order polynomial over one-second windows. Instead, Toledo et al. (2007) apply a locally weighted regression technique on the trajectory function to derive instantaneous speeds and accelerations in the central point of the moving window. As instantaneous rather than average values are returned, attention must be paid to the issue of internal consistency. For differentiation and smoothing operations, it is worth noting that the technique operates both functions at the same time.

Xin et al. (2008) employed a bilevel optimization structure that attempts to minimize measurement errors and preserve internal consistency in estimating speeds and accelerations from position data (video recordings). They attained better internal consistency results compared to locally weighted regression.

It is worth noting that none of the techniques mentioned addresses the problem of platoon consistency. To tackle this problem, Punzo et al. (2005) applied a nonstationary Kalman filter to DGPS data, introducing intervehicle spacing in model and measurement equations. The results are compared to those from the applications of a Butterworth filter to $\hat{v}_n(t)$ and of a locally weighted regression technique to $\hat{s}_n(t)$. Comparisons revealed the significant inconsistency of the trajectories filtered by the two last methods against those by Kalman filtering.

Earlier, Kalman filtering was applied in a different form to position data by Ervin et al. (1991), who designed a stationary filter to process (x,y) coordinates and yaw. They report results of application on noisy synthetic data and do not mention the consistency of the estimated trajectory. Subsequently, Ma and Andreasson (2006) applied a Kalman smoother first to displacements and speeds of an instrumented vehicle and then to intervehicle spacing from an observed vehicle (measured by a laser beam scan). However, due to the particular experimental setting, no information about platoon consistency of estimates could be derived.

Finally, from the partial information available from the NGSIM project, currently the largest available source of trajectory data, it is possible to question the processing of data. First, positions of vehicles were captured from videos and reported in a global geographical reference system. The resulting coordinates were then projected into a local coordinates system having one axis named *LocalY* aligned with the road. The speed was to be calculated from *LocalY* data using a locally weighted regression. The main difficulty with these techniques is assessing and comparing their performances. This is mainly due to the lack of an established method to quantify and measure the accuracy of trajectory data and to the lack of knowledge about the impact of such accuracy on the application results.

A recent attempt to define a quantitative methodology to inspect the accuracy of trajectory data was carried out by Punzo et al. (2011). They designed a methodology that involves an analysis of jerks, a consistency analysis, and a spectral analysis that led to the definition of a number of statistics useful to quantify data accuracy. The method was then applied to the complete set of NGSIM databases. We report below the final recommendations of the authors about the processing of NGSIM trajectory data.

Errors occurred in stopped-vehicle conditions and in the initial and final phases of trajectories. Errors in field velocity at low speeds, in particular, are due to the apparent practice of reversing the sign of negative speeds in the process of estimating vehicle velocity from *LocalY*. These considerations suggest carefully handling of values in the velocity and acceleration fields of NGSIM data sets and, when possible, directly estimating speeds and accelerations from *LocalY* data.

In light of the reasoning throughout Punzo's paper, it seems that the most effective technique would be smoothing the *LocalY* data and then differentiating to speeds (when smoothing *LocalY*, particular care should be devoted to cut-off measurement errors arising at vehicle stops). In order to dampen the residual noise, speeds should be smoothed by applying whatever filter does not introduce a bias on the space traveled (before differentiating to accelerations and smoothing again). This process would guarantee smooth speeds and accelerations while preserving the internal and platoon consistencies of trajectories.

3.3 TRAFFIC STATE ESTIMATION AND DATA FUSION TECHNIQUES

For many on-line applications, an accurate estimate of the current traffic state (speeds, flows, and densities) is required. For many short-term prediction models, these estimates serve as the initial conditions of the forecast, and the quality of the resulting prediction is directly related to the quality of the estimate. For off-line applications, improving the quality of the raw data by applying estimation and fusion techniques (noise reduction, outlier detection, etc.) is important.

This section is in large part based on chapters in the thesis of Ou (2011), Yuan (2013), and Schreiter (2013) and provides an overview of the state estimation and data fusion techniques, focusing on enhancing macroscopic traffic data. By and large, these techniques range from purely statistical procedures to methods that rely heavily on theory and assumptions stemming from our knowledge of traffic flow dynamics (theory- or model-based techniques). Before discussing techniques, we will briefly discuss preprocessing of macroscopic traffic data.

3.3.1 Short note about traffic variables and traffic states

We will now describe the main variables used to characterize traffic conditions and their links. Basically, the traffic phenomenon is the macroscopic expression of a collection of individual evolutions of vehicles in the (x, y, t) space. To characterize traffic as a whole, flow and density data are regularly used.

- Flow is the number of cars passing a certain position x during a given period Δt. It is typically expressed as vehicles per hour or minute.
- Density is the number of cars present at a certain instant t over a certain length of road Δx. It is typically expressed as vehicles per kilometer or meter.

Figure 3.1 illustrates the physical meanings of both variables. The most common measurement device for traffic is the loop detector described in Section 2.2.2.1 that functions as a point detector. Therefore, strictly

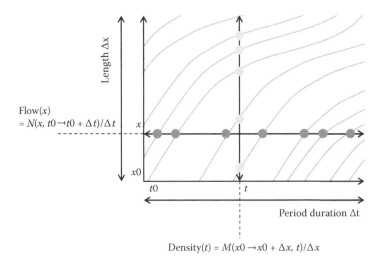

Figure 3.1 Measurement of flow and density from individual trajectories.

speaking, it is impossible to measure density. Because of that, a proxy known as *occupancy* is generally used. Occupancy is the percentage of time during which the loop is occupied by a vehicle. Note that intrinsically occupancy is not linearly related to density.

The reader is referred to Chapter 2 of traffic flow monograph the Gartner et al., (2001) for complete explanations of the various macroscopic variables of traffic. We encourage the reader to understand the differences between spatial and temporal mean speeds. Indeed, fluid mechanics studies teach us that, taken at the limit, the flow and density variables are linked through the spatial mean speed as synthesized in the following equation:

$$Q(x,t) = K(x,t) \times V(x,t) \tag{3.1}$$

Note that V in this equation is the spatial mean speed and not the temporal mean speed. The variables are linked in pairs by the fundamental relationship of traffic. Figure 3.2 shows two popular representations of a fundamental diagram realized from a typical highway. One can clearly distinguish the fluid part (Figure 3.2a), where the flow increases when the density increases, and the congested part, where the flow decreases when density increases. The same traffic states are illustrated in Figure 3.2b. In the upper part, the speed is relatively high (above 50 km/h) and the regime is fluid. The maximum flow (also called capacity) is around 4000 vehicles/h.

3.3.2 Data preprocessing and check algorithm

In many countries, macroscopic traffic data are collected from sensors such as loop detectors, radar or infrared sensors, and cameras. Unreliable measurements and disturbances impact the raw data from these sensors. These raw data typically exhibit both random errors (noise) and structural errors (bias). Although data assimilation techniques (such as the Kalman filter) are suitable to correct random errors, they typically fail to correct structural errors that may hinder further analyses. A natural question is how to eliminate structural errors in observation data.

In this section, we present a speed bias correction algorithm as one technique for dealing with the foregoing problem. This algorithm is one of the main components in the CHECK (correction algorithm for heterogeneous traffic data based on shock wave and kinematic wave theory) algorithm (Van Lint et al., 2009). It typically corrects the second (structural) category of errors in macroscopic traffic data. Specifically, it corrects biased speeds inherited from dual loop detectors, based on notions from the first-order traffic flow theory and empirical flow-density relationships.

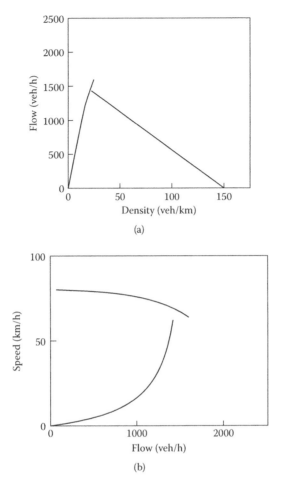

Figure 3.2 Fundamental relationship according to Wu (2002). a. Flow versus density.
b. Speed versus flow.

3.3.2.1 Problem analysis and background

In many countries such as the United States, United Kingdom, Netherlands, France, Germany, Spain, and Italy, traffic network monitoring is largely based on stationary loop detectors. Single loops (mainly implemented in the United States and in urban environments) provide flow and occupancy information that in turn may be used to estimate speed (using average vehicle length). However, these estimates are quite noisy due to unobserved variables (vehicle length, traffic density) and measurement errors.

Many researchers achieved better estimates of speed from single loop detectors (Dailey, 1999; Coifman, 2001; Jain and Coifman, 2005; Coifman and Kim, 2009). Along with flow and occupancy data, dual loops also collect (average) speeds. The Netherlands freeway network, for example, is monitored by dual loops located about every 500 m on the western part of the network. Of the available data from this dual loop system (named *MoniCa*), 5 to 10% on average is missing or otherwise deemed unreliable (Van Lint and Hoogendoorn, 2009). Similar traffic monitoring systems equipped with inductance loops can be found in England (Highway Agency, 2012) and in Germany (Schönhof and Helbing, 2007).

Locally measured and averaged speeds are not necessarily representative for spatial properties of traffic streams for two reasons. First, a local average speed at some cross section x_i over a time period ΔT is equal to the space mean speed v_M on a road section M (including x_i) over the time period ΔT only when traffic conditions are homogeneous and stationary over section m during ΔT (which does not mean that all vehicles drive at the same speed). Second, the former holds only if speeds of passing vehicles are averaged harmonically (and not arithmetically) with $v_L = \Sigma v_i/N$, where N depicts the number of observations in the period ΔT. The latter average (often called time mean speed) is biased due to an overrepresentation of faster vehicles in a time sample. In contrast, the harmonic mean speed

$$v_H = \frac{N}{\sum_{i=1}^{N} \frac{1}{v_i}} \tag{3.2}$$

essentially represents the reciprocal value of average slowness $1/v_i$, i.e., the average time per unit space each vehicle spends passing the detector. Saving for the error due to the assumption of homogeneous and stationary traffic, this average is an unbiased estimator of the space mean speed. The equivalence can be easily proved by Edie's definition of traffic variables (1965) and the work of Leutzbach (1987). The relationship between space mean speed (v_m) and local arithmetic time mean speed (v_L) can be analytically expressed by the following equation (Leutzbach, 1987):

$$v_L = v_M + \frac{\sigma_M^2}{v_M} \tag{3.3}$$

where σ_M^2 denotes instantaneous speed variance. First, the implication of Equation (3.3) is that time mean speed is always equal to or larger than space mean speed. This is a structural difference proportional to instantaneous speed variance.

Second, the effect of arithmetic time averaging cannot be undone afterward using Equation (3.3). There is no straightforward (analytical) formula to derive space mean speeds from time mean speeds directly, since the bias depends on a second unknown quantity known as instantaneous speed variance.

Many empirical studies revealed that the bias (σ_M^2/v_M) is significant and may result in speed overestimation up to 25% in congested conditions and even larger errors in travel times or densities derived or estimated from these speeds (Stipdonk et al., 2008; Stipdonk, 2013; Lindveld and Thijs, 1999; Van Lint and Van der Zijpp, 2003). A typical result from using arithmetic time averaged speeds in estimating densities (via $k = q/v_L$) is that the estimated traffic states $\{q,k\}$ (correct flow with biased density) do not lie on a semilinear line in the congested branch of the flow density diagram. Instead, they scatter widely below this line, resulting in a P-shaped distortion of the estimated fundamental $(q - k)$ diagram (Stipdonk et al., 2008); the congested phase line looks like a convex curve as the upper part of the letter P.

This effect has also been analytically proved by Stipdonk (2013). Additionally, the biased speeds may cause underestimates of route travel times (Lindveld and Thijs, 1999; Van Lint and Van der Zijpp, 2003). These biased observations provide useless information for traffic state estimation and may lead to inaccurate estimations of traffic states (e.g., densities).

Clearly, in the case of monitoring (e.g., by MoniCa in the Netherlands and the British and German highway monitoring systems) that collects time-averaged speeds, there is a need for algorithms and tools capable of correcting the inherent bias. The next subsection will first overview the approaches proposed in the literature, after which a new algorithm based on a flow density diagram and traffic flow theory is presented.

3.3.2.2 Overview of speed bias correction algorithms

Based on Equation (3.3), there are two ways of correcting the bias. One could consider the bias term $B = \frac{\sigma_M^2}{v_M}$ as a whole entity, that is,

$$v_M = v_L - B \tag{3.4}$$

where B may be a constant or some function of quantities that are available (e.g., $B = B(q,k,v,\ldots)$). One can also solve v_M from Equation (3.3), which gives us

$$v_M = \frac{1}{2}\left(v_L + \sqrt{v_L^2 - 4\sigma_M^2}\right), \sigma_M \leq \frac{1}{2}v_L \tag{3.5}$$

If the instantaneous speed variance (σ_M^2) is known or estimated, the space mean speed can be deduced accordingly. In the following, we will discuss previous research on these two topics.

The simplest method to correct speed bias is to describe the space mean speed v_M as a function of the time mean speed v_L. Essentially, with Equation (3.5), we can consider the speed variance (standard deviation) equal to a certain percentage of the time mean speed, that is, $\sigma_M = \beta \sigma_L$, with β in the order of 0.05 to 0.3 [5 to 30%, according to Lindveld and Thijs (1999)]. They also illustrated that β is not a constant, but rather is a function of traffic flow and speed. This relationship is derived using empirical data in which the ratio of standard deviation over mean speed increases with decreasing mean speed.

Van Lint (2004) proposed two other methods for estimating instantaneous speed variance in Equation (3.5) on the basis of empirical individual data collected along the A9 freeway in the Netherlands in October 1994. Figure 3.3a illustrates these data and shows two distinct regimes. Thus, these data can be fitted by a bilinear function (solid line). For low (time mean) speeds (i.e., in congestion) speed variance appears constant (although noisy) with a mean on the order of 5^2 (m/s)^2. Above 80 km/h (i.e., in free flow conditions), speed variance seems to increase steeply with time mean speeds because vehicles in congestion are constrained by one another in their choices of speeds.

The predominant cause of speed variation over space will be (collective) acceleration and deceleration due to passing stop-and-go waves. Under free flow conditions, speed variance will logically increase due to (1) traffic heterogeneity (trucks versus passenger cars) and (2) the decrease in sample sizes (number of vehicles per space and time unit) arising from decreasing density (and higher speeds).

Figure 3.3b shows a second scatter plot of instantaneous speed variance and local density k_L, which is flow divided by time mean speed q/v_L. The second approach uses an approximation of this relationship, illustrated by the dashed line in the figure. This method exploits the time series relations between the instantaneous speed variance and the variance of time mean speed. Note that this procedure contains three parameters (P is the size of consecutive time windows, k_L^{cri} denotes the critical local density, and γ is a scaling parameter) that must be tuned with respect to site-specific characteristics. More details can be found in Van Lint (2004).

Recently, Soriguera and Robusté (2011) developed a method to estimate speed variance based on probabilistic theory. However, this method requires the input of speed stratification (vehicle counts over defined speed ranges), which is commonly available from Spanish traffic monitoring systems but not elsewhere. Therefore, the application of this method is rather limited.

The methods outlined above essentially correct speed bias based on local relationships found in the available detector data. Alternatively, one could estimate speed bias also on spatiotemporal relationships found in the data, using, for example, the fundamental diagram, that relates average flow (a local quantity) to average density (a spatial quantity). Jain and Coifman (2005) validated speed estimates from single loops using flow–occupancy relationships. Coifman (2002) also exploits basic traffic flow theory and spatiotemporal propagation characteristics of traffic patterns to estimate

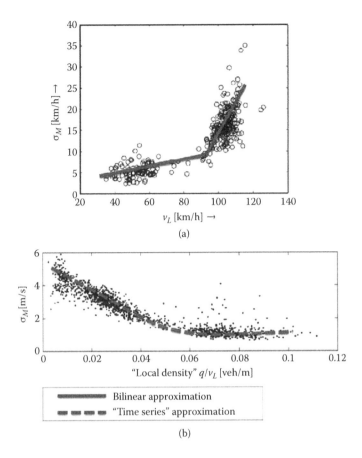

Figure 3.3 (a) Scatter plot of instantaneous speed variance versus time mean speed based on individual data collected at a single detector along the A9 on October 25, 1994, and bilinear approximation between the two quantities. (b) Scatter plot of instantaneous variance versus local density and time series approximation. (*Source:* Yuan, Y. 2012. Lagrangian multiclass traffic state estimation. PhD thesis. Delft University of Technology. With permission.)

link travel time using local dual loop data. Although these methods do not apply to the problem of estimating the bias in time mean speed directly, we are motivated to seek an alternative method to solve the problem.

3.3.2.3 Correction algorithm based on flow density relations

According to the kinematic wave theory (Lighthill and Whitham, 1955; Richards, 1956), perturbations (low speeds, high densities) propagate through a traffic stream at a speed equal to df/dk, with $q = f(k)$ depicting an equilibrium relationship (fundamental diagram) between average flow and density. This characteristic speed is typically positive in free flow and negative in congested conditions.

According to some authors (Newell, 1993; Daganzo, 1994; Kerner and Rehborn, 1997; Windover and Cassidy, 2001; Treiber and Helbing, 2002; Van Lint and Hoogendoorn, 2009), a simple and still reasonable approximation is to assume only two main characteristic speeds, one for congested traffic and one for free-flowing traffic, respectively. This results in a triangular flow density relationship:

$$q = f(k) = \begin{cases} v_{\text{free}} \cdot k, & k \le k_{\text{cri}} \\ q_{\text{cap}} + v_{\text{cong}} \cdot (k - k_{\text{cri}}), & otherwise \end{cases} \qquad (3.6)$$

where q_{cap} is the capacity flow and v_{free} and v_{cong}, respectively, depict the characteristic propagation speed in free flow and congested conditions. Note that v_{cong} is often parameterized with $v_{\text{cong}} = -q_{\text{cap}}/(k_{\text{jam}} - k_{\text{cri}})$, where k_{jam} and k_{cri} depict the jam density and critical density, respectively.

Figure 3.4 presents a typical speed contour taken from a Netherlands freeway in which the approximate constant characteristic propagation speeds can be identified. In congestion (low speeds), for example, perturbations in a traffic stream move upstream with remarkably constant

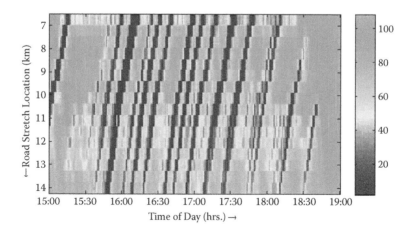

Figure 3.4 Speed contour plot measured on southbound A13 freeway between Delft North and Rotterdam Airport, Netherlands. (*Source:* Yuan, Y. 2012. Lagrangian multiclass traffic state estimation. PhD thesis. Delft University of Technology. With permission.)

speeds, illustrated by the thick dark stop-and-go waves (low speed areas) in Figure 3.4. These waves propagate upstream over the entire freeway stretch of 8 km. Note that inside the speed waves the individual vehicle speeds are not uniform.

Using the same detector data, one would expect these phenomena to translate into traffic states on a straight line in the $q - k$ diagram. As mentioned above and in Stipdonk et al. (2008) and Stipdonk (2013), the straight or semistraight line for congested traffic is not observed when we use local density $k_L = q/v_L$ as a proxy for true (but unobserved) density k. Instead, in that case, we see a P-shaped distortion attributed to the nonzero bias term σ_M^2/v_M in Equation (3.3). We can, however, with the assumption of an approximately straight congested branch of the fundamental diagram, estimate and correct this error in density and, as a result, correct the bias in speed.

To this end, a few more conditions must be met. First, if we assume that the true traffic state $\{q,k\}$ in congestion lies on a straight line with a slope equal to the propagation speed v_{cong}, we are required to estimate this parameter. For instance, the estimation can be based on spatiotemporal plots (Figure 3.4) via image processing techniques (Schreiter et al., 2010). Second, we must assume that the measured flows are unbiased, in which case correcting $k_L (= q/v_L)$ boils down to estimating the error in v_L, which of course equals σ_M^2/v_M. Figure 3.5 illustrates the basic correction principle.

The essence of this method lies in considering the bias term as a whole entity [B; refer to Equation (3.4)] to correct local time mean speed, based on notions from the traffic flow theory and traffic propagation characteristics. The detailed working principle is described below, followed by a procedure schematic (Figure 3.6).

Data input (first step in Figure 3.6) — The input data are obtained from the monitoring system in the pattern of a spatiotemporal $(x - t)$ matrix. They are designated raw observed flow q_{input} and speed v_L. Meanwhile, the

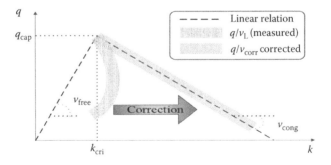

Figure 3.5 Correction principle. (*Source:* Yuan, Y. 2012. Lagrangian multiclass traffic state estimation. PhD thesis. Delft University of Technology. With permission.)

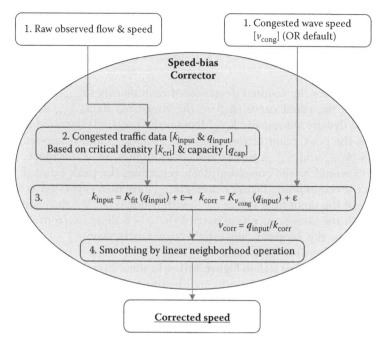

Figure 3.6 Speed-bias correction algorithm. (*Source:* Yuan, Y. 2012. Lagrangian multi-class traffic state estimation. PhD thesis. Delft University of Technology. With permission.)

propagation characteristic speed v_{cong} must be determined as the slope of the congested flow line used in this algorithm. This parameter can be estimated based on historical data (e.g., speed contour plot) or set as a default (e.g., 18 km/h as used in many applications).

Data selection (second step in Figure 3.6) — Based on local speed (v_L) and flow (q_{input}) data, the density data are generated via $k_{input} = q_{input}/v_L$. As the focus is on the congested traffic, the free flow state and the congested state must be distinguished. In macroscopic traffic flow models, critical density on a road segment discriminates free flow from congested flow conditions. Here, the critical density value (k_{cri}) and/or the capacity value (q_{cap}) that relate to the peak of a $q - k$ diagram serve the same purpose. The k_{cri} and q_{cap} values can be easily estimated from the input data. The data sets in terms of flow and density that meet the condition $k_{input} > k_{cri}$ are categorized as congested state data.

Additionally, we need to ensure that the data selected from congested conditions represent stable traffic states, that is, traffic states that are likely to lie on the (linear) congested branch of the fundamental diagram. This implies we cannot select detector data located downstream (of a bottleneck). At locations where congestion is resolving, flows are typically far below capacity with large spacing (low density) due to accelerating vehicles.

Density-wise correction (third step in Figure 3.6) — This step is the core of the correction algorithm. The mapping of this operation is expressed by:

$$k_{\text{input}} = K_{\text{fit}}(q_{\text{input}}) + \varepsilon \rightarrow k_{\text{corr}} = K_{v_{\text{cong}}}(q_{\text{input}}) + \varepsilon \tag{3.7}$$

where ε denotes the residual deviation of each density (k_{input}) scatter point away from the fitted curve (K_{fit}) of the congested data, k_{corr} denotes the corrected density values, and $K_{v_{\text{cong}}}$ denotes the targeted congested branch.

Given the peak point (fitting origin: k_{cri} and q_{cap}) and congested traffic scatters $(k_{\text{input}}$ and $q_{\text{input}})$, fit function (K_{fit}) applies. In $q - k$ space, the fit function would fit the congested data, regarding the peak point $(k_{\text{cri}}$ and $q_{\text{cap}})$ as the origin point. Density values (k_{input}) are then expressed as a function K_{fit} of the input flow values q_{input} with residual (ε). We then shift these scatters in the congested phase part of the $q - k$ diagram from the fitted curve K_{fit} to the targeted congested characteristic line $(K_{v_{\text{cong}}})$. The fitted curve can be linear or nonlinear.

Postprocessing (last step in Figure 3.6) — In some cases, samples with high speeds (e.g., exceeding 120 km/h) are identified as erroneous. Moreover, in the speed space, the corrected speed values (v_{corr}) of the congested state may contain discontinuity with respect to those in free flow state, so a linear smoothing can be used. The smoothing is a neighborhood operation already proposed in Treiber and Helbing (2002a) and Van Lint and Hoogendoorn (2009). The operation for each point is conducted within a small spatio-temporal area based on the values of its neighboring points to overcome the discontinuity problem. Finally, the corrected speed data are obtained.

3.3.3 Kalman filters and their variations

In this subsection, we briefly consider one of the most applied approaches to state estimation and data fusion, namely, Kalman filters. We first introduce the concepts, after which we will show some of the applications of data fusion applied to traffic operations.

3.3.3.1 Overview of common Kalman filtering approaches

The most widely utilized data assimilation technique applied to traffic state estimation problems is the Kalman filter (KF) and/or its many variations (extended and unscented KFs). Kalman filters for state estimation and data assimilation utilize the ability of many analytical traffic models to be expressed in state space form, that is:

$$x_t = f(x_{t-1}, u_t) + w_k \tag{3.8}$$

$$y_t = h(x_t) + v_t \tag{3.9}$$

Here, t depicts discrete time steps of duration Δt seconds. Equation (3.8) depicts the process equation also known as state transition equation, which describes the dynamics of state x_t (e.g., density and/or speed) as a function of x_{t-1} and external disturbances u_t (for example, traffic demand at network boundaries) plus an error term w_t reflecting errors in the process model (misspecification, process noises).

Equation (3.9) depicts the observation equation also known as measurement equation h, which relates the system state to measurement y_t. The error term depicts errors in either the measurement model h and/or the measures themselves. The fundamental diagram of traffic flow $Q = Q^{eq}(K)$, or $U = U^{eq}(K)$, relating speed or flow to density, is a good example of such a measurement equation. Q and U represent the flow and speed measurements from loop detectors, and K represents the density variables that need to be estimated.

If the above equations represent a linear dynamic system, f and h are the linear operators that can be expressed by matrices F_t and H_t, respectively. As a result, the following equations can be derived:

$$x_t = F_t x_{t-1} + B_t u_t + w_t \tag{3.10}$$

$$y_t = H_t(x_t) + v_t \tag{3.11}$$

where w_t is assumed to be drawn from a zero mean multivariate normal distribution with covariance $\mathrm{Cov}W_t$, and v_t is assumed to be a zero mean Gaussian white noise with covariance $\mathrm{Cov}W_t$. Note that the initial state and the noise vectors at each time step are all assumed to be mutually independent.

In what follows, let the notation $\hat{x}_{(n|m)}$ represent the estimate of x at time n given observations up to and including at time m. The state filter is represented by two variables: $\hat{x}_{(t|t)}$ is the a posteriori state estimate at time t given observations up to and including at time t; $\hat{P}_{(t|t)}$ is the a posteriori error covariance matrix for the state estimate at time t given observations up to and including at time t. The initial conditions are:

$$\hat{x}_{(0|0)} = \hat{x}_0, \ \hat{P}_{(0|0)} = \hat{P}_0 \tag{3.12}$$

With the initial conditions, a Kalman filter is iteratively executed in two distinct steps:

State prediction:

$$\hat{x}_{(t|t-1)} = F_t \hat{x}_{(t-1|t-1)} + B_t u_t \tag{3.13}$$

State correction:

$$\hat{x}_{(t|t)} = \hat{x}_{(t|t-1)} + K_t (y_t - H_t \hat{x}_{(t|t-1)}) \tag{3.14}$$

The so-called Kalman gain K_k in Equation (3.14) is computed to make the Kalman filter an optimal estimator in terms of least squares:

$$K_t = \hat{P}_{(t|t-1)} H_t^T S^{-1} \tag{3.15}$$

where S_t^{-1} is called the innovation covariance or residual. It is worth noting that the computation of the inverse of the matrix may be very time-consuming. The details can be found in Kalman (1960). The formally can be informally stated as:

$$K_k = \frac{uncertainty\ process\ model}{undercertainty\ observation\ model} \\ \times sensitivity\ obs.\ model\ to\ state\ variables \tag{3.16}$$

This implies that (1) the more uncertain the data, the more weight is put on the model predictions and vice versa; and (2) that the KF adjusts x_t proportionally to the sensitivity of the observation model to changes in the state variables. For example, under free flow conditions, the relationship between traffic density and speed is very weak, which would imply only small corrections in state variables (x_t) even if the speeds measured by sensors (y_t) differ largely from those predicted by the observation model $(H_t \hat{x}_{(t|t-1)})$.

This intuitive structure can be easily explained to traffic operators and professionals using such state estimation tools. Moreover, the same process and observation model can be used subsequently for prediction and control purposes if proper predictions of the boundary conditions (traffic demand, turn fractions, and capacity constraints) and estimates of the model parameters are available.

When the dynamic system is nonlinear, which is typically the case in traffic, f and h cannot be expressed by matrices F_t and H_t. However, the system can be linearized by computing a matrix of partial derivatives (Jacobean) around the current estimate. This linearization forms the core of the so-called extended Kalman filter. Unlike the standard Kalman filter, the extended version is not an optimal estimator when the process model or observation model is not linear. If the initial estimates of the state or the process model are not correct, the filter may quickly diverge due to linearization.

An improvement to the extended Kalman filter (EKF) led to the development of the unscented Kalman filter (UKF), which is also nonlinear. Its probability density is approximated by a nonlinear transformation of a random variable, leading to more accurate results than the first-order Taylor expansion of the nonlinear functions in the EKF. The approximation utilizes a set of sample points that guarantee accuracy with the posterior mean and covariance to the second order for any nonlinearity. In addition, unlike the EKF, there is no need in the UKF to calculate the Jacobian. However, there is a demand for computing many sample points.

Both EKF and UKF assume Gaussian distributions of the process of the noise in Equation (3.10) and observation in Equation (3.11). These methods fail when the distributions are heavily skewed, bimodal, or multimodal. To handle any arbitrary distribution, particle filters are proposed as an alternatives to the EKF and UKF for non-Gaussian distributions. Particle filters are simulation-based techniques that are able to approach Bayesian optimal estimates with sufficient samples.

3.3.3.2 Application of Kalman filters to traffic data fusion

Many data fusion methods for traffic state estimation use the EKF and its variations as the data assimilation technique. They differ mainly in data input, data output, traffic models, and assumptions. Gazis and Knapp (1971) use time series flow and speed from loop detectors to estimate traffic density. Basic physical laws are commonly used to approximate travel time on a road section, and then Kalman filters are applied to combine data and the model. Szeto and Gazis (1972) estimate traffic density between the two consecutive loops by fusing aggregated loop speeds and flows.

The traffic model is based on the vehicle conservation law and speed–density relation. Nahi and Trivedi (1973) also used loop flows and speeds as input data. This method contains a simpler traffic model that simply employs the conservation law and can estimate both density and speed. Ghosh and Knapp (1978) approximated the space speed over two consecutive loops by simply averaging speeds from the two. As a result, a linear state model can be achieved by exploiting the conservation law.

Using a different input, another contribution is reported in Kurkjian et al. (1980). They managed to use loop flow and occupancy to estimate traffic density. The traffic models used in the above methods are first-order macroscopic models and most of them employ vehicle conservations laws and do not consider speed–density relations.

Since the end of 1970s, more advanced traffic models came into use. Almost simultaneously, Willsky et al. (1980) and Cremer and Papageorgiou (1981) combined a second-order macroscopic traffic model and Kalman filter to estimate traffic states (speed, flow, and density) by using loop speed and flow. Willsky et al. used the Payne model. Following a similar method, Kohan and Bortoff (1998) proposed a nonlinear sliding mode observer when combining a Kalman filter and a second-order macroscopic model.

Also based on the Payne model, Meier and Wehlan (2001) proposed a new scheme called section-wise modeling of traffic flow that helped approximate the boundary variables between the sections. Exploiting these methods to the extent possible, Wang and Papageorgiou (2005) proposed a general approach to the real-time estimation of the complete traffic state on freeway stretches based on an EKF and a second-order traffic model. This method allows the unknown but important parameters such as free speed, critical density, and exponent to be on-line estimated. Their further study and applications were shown in later publications as Wang et al. (2006), Wang et al. (2008), and Wang et al. (2009).

In addition to loop data, other types of data also can be fused by employing a Kalman filter and its variations. Chu et al. (2005) estimated traffic density and travel time by fusing loop flow and probe car travel times over a section. In this method, the traffic within a section is assumed to be homogeneous and probe vehicles provide travel times over the section that are used as measurements. The assimilation technique is adaptive Kalman filtering. Herrera and Bayen (2007) estimated density by fusing loop data and vehicle trajectories from mobile sensors and employed a first-order traffic model as a process model. Loop flow is used as Eulerian measurement and the vehicle trajectory as Lagrangian measurement from which local density is computed.

Figure 3.7 shows the performance of an EKF. The results become smoother after applying the filter.

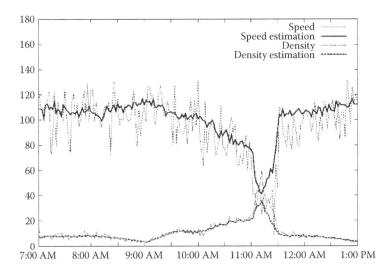

Figure 3.7 Estimation resulting from use of extended Kalman filter for loop speed and flow data with second-order traffic model. (*Source:* Adapted from Wang, Y. and Papageorgiou, M. 2005. *Transportation Research Part B,* 39(2), 141–167. With permission.)

Within the framework of particle filters, Mihaylova et al. (2007) used loop speed and flow to estimate the traffic states (speed, flow, and density). A second-order macroscopic traffic model was employed to establish process equations and observation equations.

Cheng et al. (2006) also used particle filters to estimate traffic states from cell phone network data. In wireless communication networks, each base station is responsible for the service within a certain area known as a cell. When a cell phone moves from one service cell to another, the communication service for the cell phone is handed over from one base station to another. The base station records the switching times so that travel time for a vehicle can be known. Cheng et al. used the hand-off technique to determine section speed and traffic flow with known probe penetration rates. A first-order and second-order traffic model are used for comparison. The estimated states are flow and speed.

Hegyi et al. (2006) investigated a comparison of an EKF and a UKF for traffic state estimation using simulated loop data. They found that although the UKF had the advantage of propagating the state noise distribution with higher precision than the EKF, its performance was nearly equal (slightly better) than that of the EKF. They also found that fewer detectors resulted in larger state estimation errors, but produced no effect on parameter estimation errors.

Ou (2011) remarked about the performances of particle filters. The true posterior probability distribution can be well approximated only when there are enough particles. Therefore, if the assumptions for Kalman filters can be guaranteed, no particle filters can outperform them. In addition, computational costs for particle filters are far higher than those for Kalman filters and their variations.

3.3.4 Alternatives to Kalman filters

Alternative approaches to Kalman filterbased state estimation and data fusion have been put forward in the literature. A successful algorithm was proposed by Treiber and Helbing (2002). The filter was originally designed for processing single data sources and reconstructing spatiotemporal traffic plots and is based on the spatiotemporal characteristics of traffic flow, that is, perturbations in traffic travel along "characteristics." For explanations of characteristics, see Chapter 5 of Gartner et al. (2001).

The traffic state estimator discussed in the remainder of this section is the adaptive smoothing method (ASM) of Treiber and Helbing (2002b). Since its conception, it has been generalized to multiple data sources (Van Lint and Hoogendoorn, 2010; Treiber et al., 2011) and used in various applications (Van Lint, 2010; Kesting and Treiber, 2008). The ASM estimates traffic state based on spatiotemporal data. Its methodology is presented in this section.

Conventional implementations require several minutes to estimate the traffic state. After a reformulation, this problem can be solved by a cross-correlation operation that operates on the whole spatiotemporal data matrix at once. This methodology was further improved by solving the cross correlation with the fast Fourier transform from the field of signal processing. The three implementations of the ASM were tested with data from a Netherlands freeway. The results show improvements of computation time up to two orders of magnitude and run within a few seconds. The proposed implementations can therefore replace the conventional implementation in practical applications.

The ASM takes speed data $v^{\text{raw}}(x,t)$ as input, observed at locations $x \in X^{\text{raw}}$ at times $t \in T^{\text{raw}}$. A second spatiotemporal traffic data variable is optional. For instance, the flow observed at the same points as the speed is used, but other macroscopic quantities such as traffic density can also be used. In the remainder of this section, the z symbol refers to any macroscopic traffic quantity, whereas v specifically denotes speed.

The output of the ASM is a continuous spatiotemporal variable designated z^{out}. To solve the ASM numerically, however, the filtered map is discretized at locations X^{out} and times T^{out}. Usually, this underlying space–time grid is chosen to be equidistant with resolution Δx^{out} and Δt^{out}, respectively.

The calculation of the output map is based on kinematic wave theory. Depending on the underlying traffic regime, the characteristics of traffic travel with a certain wave speed over space and time. Traffic is to be assumed in one of two regimes. Each regime has one typical wave speed with which the characteristics travel. In free flow, this wave speed is approximately c_{free} = 80 Km/h and in congestion is approximately c_{cong} = 18 km/h. The exact values of these two calibration parameters can be determined by applying the wave speed estimator in Equation (3.32).

The data map z^{raw} is nonlinearly transformed into a smooth map z^{out}, whose elements represent a weighted sum of smoothed elements of both traffic regimes:

$$z^{\text{out}}(x,t) = w(x,t) \cdot z^{\text{cong}}(x,t) + [1 - w(x,t)] \cdot z^{\text{free}}(x,t) \tag{3.17}$$

The intermediate functions z^{cong} and z^{free} represent the traffic in congested and in free flow conditions, respectively. The weighting factor w depends on the underlying traffic regimes. The congested function z^{cong} is defined by:

$$z^{\text{cong}}(x,t) = \frac{1}{n^{\text{cong}}(x,t)} \sum_{x_i} \sum_{t_j} \varphi^{\text{cong}}(x_i - x, t_j - t) \cdot z^{\text{raw}}(x_i, t_j) \tag{3.18}$$

with the normalization factor

$$n^{cong}(x,t) = \sum_{x_i} \sum_{t_j} \varphi^{cong}(x_i - x, t_j - t) \qquad (3.19)$$

whereby the sums cover all date locations $x_i \in X^{raw}$ and data times $t_j \in T^{raw}$. The smoothing kernel

$$\varphi^{cong}(x,t) = \exp\left(-\frac{|x|}{\sigma} - \frac{\left| t - \frac{x}{c_{cong}} \right|}{\tau} \right) \qquad (3.20)$$

is an exponential function with the spatial parameter σ and the temporal parameter τ. The characteristic congested wave speed c_{cong} influences the skew of the kernel.

The free flow function z_{free} is similarly defined to z_{cong} [Equations (3.18) through (3.20)] with a normalization factor n^{free}, the free flow smoothing kernel φ^{free}, and the free flow wave speed c^{free}. The weighting factor w in Equation (3.19) depends on the intermediate speed functions v^{cong} and v^{free}:

$$w(x,t) = \frac{1}{2}\left[1 + \tanh\left(\frac{v_{crit} - v^*(x,t)}{\Delta v} \right) \right] \qquad (3.21)$$

With critical speed v_{crit}, transition speed range Δv and

$$v^*(x,t) = \min(v^{cong}(x,t), v^{free}(x,t)) \qquad (3.22)$$

For details about the ASM, we refer to the original paper (Treiber et al., 2011). The result z^{out} is defined in continuous space time. For numerical computations, however, the above equations are discretized. Conventional algorithms implement the double sum of z^{cong} (3.18) in a double loop.

An example of input and output data of the ASM is shown in Figure 3.8. The induction loops of the A15 eastbound near Rotterdam, Netherlands, collected speed data over a length of $X = 30$ Km and a measurement time of $T = 10$ h. Both traffic regimes, free flow (light colors) and congestion (dark colors), are present. Note that some detectors do not provide data, thus leaving holes in the data map (Figure 3.8a). The ASM smooths these data to generate a complete speed map (Figure 3.8b).

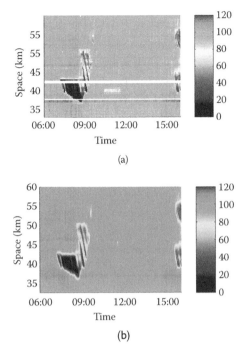

(a)

(b)

Figure 3.8 Input and output speed data of ASM. FFT = fast Fourier transform. (*Source:* Schreiter, T. 2013. Vehicles-class Specific Control of freeway Traffic, Ph.D. thesis, Delft University of Technology. With permission.)

3.4 FEATURE EXTRACTION AND PARAMETER IDENTIFICATION TECHNIQUES

The previous sections discussed Level 1 estimation and data fusion approaches for microscopic and macroscopic traffic flow data. After such estimates have been acquired, other inferred information about these enhanced data are often a useful component for model calibration and validation. This process, known as feature extraction, will be discussed in this section.

Let us note that calibrating models can in fact also be seen as a form of (model-based) feature extraction. Calibration is discussed in Chapters 4 and 5. Here, we focus on features that can be more or less directly inferred from the enhanced data, e.g., by recognition of specific patterns within the data.

In the past, many approaches for dealing with feature extraction have been put forward. This section focuses on three applications: estimation of free speed distribution (microscopic), extraction shock wave speeds (macroscopic), and estimation capacity (macroscopic).

3.4.1 Free speed distribution estimation

Knowledge of free speeds on a road under given conditions is relevant for a number of reasons. For instance, the concept of free speed is an important element in many traffic flow models. Free speed distribution is an important input for so-called gas-kinetic models. Many microscopic simulation models draw the free speeds of individual vehicles from free speed distributions.

Insights into free speeds and their distributions are also important from a design view and for determining suitable traffic rules for a specific situation. For instance, elements of a network should be designed so that drivers using it can traverse the roads safely and comfortably. It is also of interest to see how desired speed distributions change under varying road, weather, and ambient conditions, and how these distributions vary for different types of travelers. Free speed of a driver population is an important design characteristic.

Generally, the free speed or desired speed of a driver–vehicle combination (simply called vehicle or driver in the ensuing text) is defined by the speed driven when the driver is not influenced by other road users. The free speed will be influenced by characteristics of the vehicle, the driver, the road, and conditions such as weather and traffic rules (speed limits). Botma (1999) describes how individual drivers choose their free speeds and discusses a behavioral model relating the free speed of a driver to a number of counteracting mental stresses to which he or she is subjected. These models have not been successful in practical application.

Estimation of free speeds and the free speed distribution is not a straightforward task. Assuming that the drivers operate in one of two states (car following or driving at free speed) suggests that considering only drivers who are driving freely will provide an unbiased estimate of free speed distribution. This is, however, not the case since drivers in relatively high free speed conditions have greater probabilities of being constrained than drivers with relatively low free speeds (Botma, 1999). Botma presented an overview of alternative free speed estimation approaches covering the following approaches:

- Estimation of free speed by considering the speed at low volumes (Transportation Research Board, 2000). This method has the drawback that low volumes generally only occur at non-peak periods (e.g., at night). The free speed distributions will differ during off-peak and peak periods due to changed ambient conditions and driver populations.
- Extrapolation toward low intensities. This method utilizes the relevant driving population but is known to be prone to errors (Botma, 1999).

- Application of simulation models. This method entails using a microscopic simulation model to establish relations between observable variables and free speed distribution (Hoban, 1980).
- Method based on Erlander's model (1971). An integral equation is developed for traffic operations on two-lane roads with free speed distribution as one of its components.

Botma (1999) concluded that all methods above present serious disadvantages, which is the reason another estimation approach has been proposed. The approach is based on the concept of censored observations using a parametric estimation approach to estimate the parameters of the free speed distribution. Speed observations are marked as censored (constrained) or uncensored (free flowing) using subjective criteria (headway and relative speed). In the remainder of this section, we will discuss the approach of Botma (1999) that entails estimating the distribution F^0 (v^0) of free speed V^0 for individual vehicle measurements collected at a cross section.

Let us assume that at a cross section x, we collected individual vehicle speeds v_i. As with the sample of headways, there are different ways to visualize a sample $\{v_i\}$. Figure 3.9 shows the empirical cumulative distribution

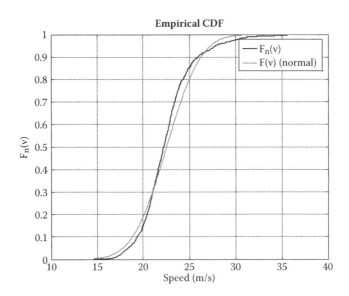

Figure 3.9 Empirical cumulative distribution function for speeds measured at Maarn measurement site. Sample mean and standard deviation are equal to 22.6 m/s and 2.90 m/s, respectively. The figure also shows normal distribution fitted to data. (*Source:* Hoogendoorn, S.P. and Van Lint, H. 2010. Contribution to CIE4831. Delft University of Technology. With permission.)

function of a headway sample collected on a two-lane rural road at the Maarn measurement site.

Similar to the headway distribution, different distribution models can be fitted to the data. In many cases, the normal distribution appears to fit speed measurements reasonably well for a sample collected during stationary traffic flow conditions.

Following and free drivers — The rationale behind the approach is the distinction between following (constrained) and free (unconstrained) drivers. The first question to be answered is how to distinguish the two. The headway can be used. However, because the minimum headway (or empty zone) is generally a distribution, putting forward a single deterministic value h^0 for which $h_i < h^0$ implies that a driver is car following while $h_i > h^0$ implies free driving is possible.

One could consider using a headway value h^* that is sufficiently large to ensure that $h_i > h^*$ implies that driver i is indeed car following. Later, however, we will show that we need a better distinction of free drivers and following drivers to derive better estimation results.

Therefore, the classifications of free and following drivers will be based on both headway and relative speed criteria, i.e., driver i is car following, if and only if:

$$h_i < h^* \quad \text{and} \quad |v_{i-1} - v_i| < \Delta v^* \tag{3.23}$$

The boundary values h^* and Δv^* are to a certain extent arbitrary, but the sensitivity of the method proposed later is relatively minor. For now, we will use $h^* = 7$ s and $\Delta v^* = 2.5$ m/s.

Figure 3.10 shows the empirical distribution functions of the speeds of the free drivers and following drivers. The former could (at a first glance) be used as an estimate for the free speed distribution.

Censored observations — An improved estimation of the free speed distribution can be acquired by noticing that the speeds of the free drivers contain information on the free speed distribution and on the speeds of the following drivers. We know that the free speed v_i^0 of driver i is larger than or equal to the observation v_i. We say that we have a right-censored observation of the free speed. Correctly including this information is particularly important, since vehicles with higher free speeds have greater probabilities of being constrained. Hence, only considering the free speeds of the free drivers will yield a bias in the estimate.

Approaches based on censored data such as that of Kaplan and Meier (1958)—also known as product limit methods—are based on classifying observations into two groups: censored and uncensored measurements. The uncensored measurements are the speed observations v_i of the

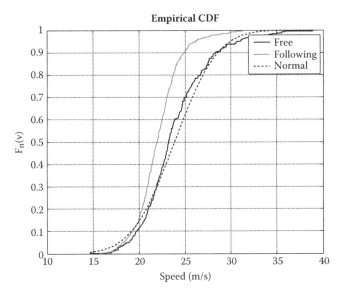

Figure 3.10 Speed distributions of free and following drivers. Note that sample mean speeds and variances of the free drivers are, respectively, equal to 23.9 m/s and 3.69 m/s. For illustration, we again fit a normal distribution for speed distribution of free drivers. (*Source:* Hoogendoorn, S.P. and Van Lint, H. 2010. Contribution to CIE4831. Delft University of Technology. With permission.)

unconstrained drivers ($\delta_i = 0$) and the censored measurements are the speed observations of the constrained drivers ($\delta_i = 1$).

3.4.1.1 Parametric Kaplan–Meier (or product limit) approach

Let us now show how we can determine the likelihood of a sample $\{v_i, \delta_i\}$. Let f_θ and F_θ, respectively, denote the probability distribution function (pdf) and the cumulative distribution function (cdf) of the free speed distribution, with unknown parameters θ (e.g., the normal distribution). When a vehicle is constrained $\delta_i = 1$, the free speed v_i^0 will be higher than the speed observation v_i. The probability of this occurring equals:

$$\Pr(V^0 > v_i) = 1 - F_\theta^0(v_i) \tag{3.24}$$

When a vehicle is driving freely ($\delta_i = 0$), the free speed v_i^0 is equal to the observed speed. The probability of this occurring is relative to:

$$f_\theta^0(v_i) \tag{3.25}$$

For an entire sample of speed measurements, the likelihood can be determined as follows:

$$L(\theta) = \prod_{i=1}^{n} (1 - F_\theta^0(v_i))^{\delta_i} \cdot (f_\theta^0(v_i))^{1-\delta_i} \tag{3.26}$$

The maximum likelihood estimate for θ can be determined by maximization of this likelihood. It is, however, generally more convenient to consider the log-likelihood:

$$\tilde{L}(\theta) = \sum_{i=1}^{n} \delta_i \log(1 - F_\theta^0(v_i)) + (1 - \delta_i)\log(f_\theta^0(v_i)) \tag{3.27}$$

Figure 3.11 shows the estimation results when applying the parametric Kaplan–Meier approach to the data collected at the Maarn site. The differences between the outcomes of the different approaches are remarkable.

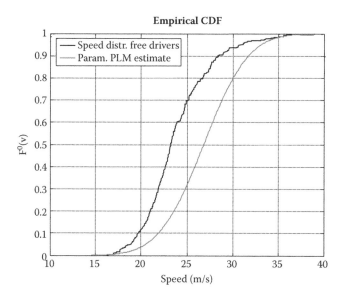

Figure 3.11 Estimation results for parametric Kaplan–Meier using normal distribution. Results for mean and standard deviation of normal distribution equal 26.80 m/s and 3.82 m/s. Note large differences of the mean and standard deviation for empirical distribution of free-flowing drivers. (*Source:* Hoogendoorn, S.P. and Van Lint, H. 2010. Contribution to CIE4831. Delft University of Technology. With permission.)

3.4.1.2 Nonparametric Kaplan–Meier approach

The parametric Kaplan–Meier approach has the disadvantage that a distribution of the free speeds needs to be determined in advance. The nonparametric approach resolves this and allows the determination of the Kaplan–Meier empirical distribution function. It is also known as the nonparametric product limit method (PLM). The first step of applying the approach is sorting the speed observations v_i. Given a certain free speed value v^0, application of the approach of Kaplan and Meier (1958) yields:

$$\hat{F}_n^0\left(v^0\right) = 1 - \prod_{j=1}^{n(v^0)} \left(\frac{n-j}{n-j+1}\right)^{\delta_j}$$

(3.28)

In this equation, $n(v^0)$ denotes the number of observations $\{v_i\}$ that are smaller than or equal to v^0 (i.e., $v_i \leq v^0$); n denotes the total number of observations. Application of the approach is left to the reader as an exercise.

3.4.2 Estimation of spatiotemporal traffic characteristics

The estimation of spatiotemporal characteristics is an important part of estimating the current traffic state. This section focuses on the estimation of shock waves and their propagation speeds, since they will be used later to calibrate the traffic state estimators.

A typical example of a shock wave occurs at stop-and-go waves. From a driver's perspective, traffic slows suddenly and drivers have to brake; after a few minutes, traffic resumes free flow conditions and drivers can continue their trips. When observing the traffic stream from above, the stop-and-go wave is visible as a region a few hundred meters long where traffic stands still. Since vehicles leave that region at its head and new vehicles enter the region at its tail, the stop-and-go wave propagates upstream. A stop-and-go wave is thus characterized by two shock waves, one at the head and one at the tail.

This pattern is visible in the printouts of spatiotemporal traffic flow data. Figure 3.12a shows multiple stop-and-go waves in a speed contour plot. The upstream shock of a stop-and-go wave is marked by the dark line. The shock wave emerges at Km 18 at 15:35 and then propagates upstream at a constant speed c of –19 Km/h. Shock waves also occur in free flow traffic, as the spatiotemporal flow plot of Figure 3.12b shows. Vehicle platoons cause regions of high flow that are surrounded by shock waves. One region in the figure is marked by a dark line.

Empirical data show that the propagation speeds of the characteristic congestion wave speed vary between –25 and –15 Km/h (Schönhof and

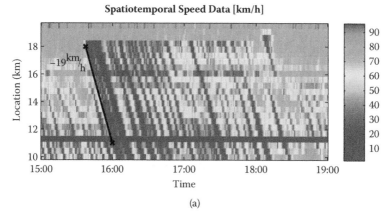

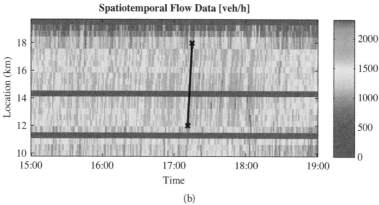

Figure 3.12 Shock waves observed in spatiotemporal freeway traffic data from A13R, Netherlands, April 2009. (a) Shock wave at upstream front of a stop-and-go wave (April 24). Shock wave as platoons of vehicles travel through light traffic (April 25). (*Source:* Schreiter, T., 2013, Vehicle class Specific Control of Freeway Traffic, Ph.D. thesis, Delft University of Technology. With permission.)

Helbing, 2009; Kerner and Rehborn, 1997; Treiber et al., 2000; Kerner, 1998; Chiabaut and Leclercq, 2011; Bertini and Leal, 2005; Graves et al., 1998; Zhang and Rice, 1999; Cassidy and Mauch, 2001). This variation can be explained with the following simple car-following law (Pipes, 1967):

$$r = r^{\min} + h^{\min} \cdot v \quad \leftrightarrow \quad v = \frac{r - r^{\min}}{h^{\min}} \tag{3.29}$$

where r denotes the average gross distance headway between vehicles, $r^{\min}$ is the distance at standstill, v represents the speed, and $h^{\min}$ is the minimum

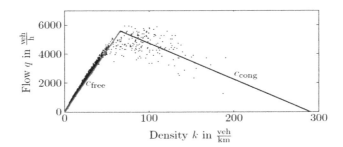

Figure 3.13 Relation between characteristic wave speeds and flow density fundamental diagram. Data from dual inductive loops at Km 55.1 on eastbound Netherlands freeway A15 on March 20, 2012. (*Source:* Schreiter, T., 2013, Vehicle class Specific Control of Freeway Traffic, Ph.D. thesis, Delft University of Technology. With permission.)

time headway. Note that this relation holds only holds at sufficiently high densities k. Since $k = \frac{1}{r}$, from Equation (3.29), it follows that:

$$v(k) = \frac{\frac{1}{k} - r^{min}}{h^{min}} \quad \rightarrow \quad q = kv(k) = \frac{1 - r^{min}k}{h^{min}} \tag{3.30}$$

showing that this car-following law leads to a linear congestion branch of the fundamental diagram expressed in flow q and density k as illustrated in Figure 3.13. Then,

$$c_{cong} = \frac{dq}{dk} = -\frac{r^{min}}{h^{min}} \tag{3.31}$$

shows the relation between c_{cong} and the car-following parameters. This relation provides a tool to show how changes in composition, road condition, and weather factors affect wave speed. For instance, a large number of trucks change the average distance at standstill r^{min} substantially, causing an increase in the speed c_{cong}. Reductions in the minimum headway h^{min} caused by changes in weather conditions, geometry, or improved visibility also lead to increased wave speed (in absolute terms). The correct estimation of the site-specific c_{cong} is thus important.

In low densities, the flow grows approximately linear with density, as the left branch of the fundamental diagram in Figure 3.13 shows. The characteristic waves speeds correspond therefore directly to the slope of the two branches of the fundamental diagram. This relationship is also used in traffic state estimation with the ASM (Treiber et al., 2011) analyzed in Section 3.3.4. The characteristic wave speeds are thus used to calibrate the fundamental diagram and traffic state estimators.

Conversely, the estimation of the characteristic shock wave speeds is possible from the flow density fundamental diagram. However, the direct estimation of the fundamental diagram from a single stationary loop detector is difficult, since the data are very noisy. An example is shown Figure 3.13 by the dots that represent data gathered at a single-lane detector. A reason for this noise is that stationary loop detectors average the data over one minute, i.e., the data are aggregated over multiple traffic states, leading to measurement errors (Laval, 2010).

On Netherlands freeways, the sensors aggregate the speed arithmetically, which leads to a systematic overestimation (Knoop et al., 2009) and therefore to a bias in the fundamental diagram. Shock wave speeds can therefore hardly be estimated accurately from single stationary detector data.

A better approach is to analyze spatiotemporal data directly as shown in Figure 3.12. The transitions of traffic states and therefore the shock waves are directly visible despite the imprecision and bias introduced by the aggregation of data. We therefore developed a method that directly analyzes spatiotemporal data and scans the resulting plot for the transition of traffic states. These shock waves are localized and their speed is averaged for each regime so that the characteristic shock wave speed is determined.

3.4.2.1 Methods for shock wave speed estimation

The most intuitive method of identifying shock waves is to manually inspect spatiotemporal traffic data plots like the ones in Figure 3.12. A shock wave is visible as the border between two areas of different colors. By measuring the spatial (Δx) and temporal (Δt) spread of that line, the propagation speed c of the corresponding shock wave speed is the slope of that line:

$$c = \frac{\Delta x}{\Delta t} \tag{3.32}$$

Methods have been developed to support this process. Zheng et al. (2011a) analyzed the speed signal of a stationary detector. A strong change of speed indicates a change of the traffic regime. If two of these changes are found within a few minutes, then a stop-and-go wave likely passed. Such a strong change is detected by applying a wavelet transform, which is a common method in signal processing. More specifically, a peak in the energy distribution of the wavelet transform indicates a strong change of speeds over a longer period. In this way, the wavelet method is robust for narrow peaks in the signal as it can be caused by detector errors.

By applying this method to multiple adjacent detectors, the shock waves are located in space and time by tracing the peaks of each wavelet signal. A linear fit of these peaks then yields the propagation speed of the stop-and-go wave. Similarly, Zheng et al. (2011b) applied the same wavelet transform to vehicle trajectories to estimate the shock wave speeds.

Another method is to estimate the propagation wave speed by matching cumulative vehicle count curves of neighboring detectors. According to Newell's car-following model (2002), the trajectories of successive vehicles in congestion are similar; they are only shifted in space and time by the wave speed. This principle was used by Chiabaut et al. (2010) by comparing successive vehicle trajectories.

Chiabaut and Leclercq (2011) applied the same principle to macroscopic data: the shift between the speed and cumulative flow counts of neighboring detectors determines the shock wave speed. This method works for single links. However, since the cumulative vehicle count curves are disrupted at non-flow-conserving nodes like on ramps, this method is not suitable for freeway sections that contain on- or off-ramps.

Treiber and Kesting (2012) analyzed the speed time series of speed data of neighboring detectors by using the cross correlation that expresses how similar two signals are. Since the time series in congestion are shifted by the characteristic shock wave speed c, the cross correlation is maximized when the downstream time series is shifted by $\Delta t = \Delta x \cdot c$ (Δx is the distance between the detectors).

However, different regimes are characterized by different wave speeds. A regime change thus leads to different signals between the detectors so that the cross-correlation approach may not work if applied to the whole signal. This method therefore requires a preselection of the data so that only data of the same traffic regime are used in the analysis. Similarly, Coifman and Wang (2005) used the cross correlation of the cumulative vehicle count and speed signals between neighboring detectors to determine the shock wave speeds in congestion.

3.4.2.2 New method for shock wave speed estimation

This section develops the methodology of the wave speed estimator (WSE; see Schreiter et al. [2010b] for details). Based on spatiotemporal speed and flow data from freeways, the WSE computes the shock wave speeds. It consists of four components, as Figure 3.14 illustrates.

The two main components originate from the field of image processing. An edge detector localizes the border between two objects. A line detector can localize straight lines. The research field of image processing is well known and widely applied in traffic science and in many other domains. Applications are, for example, license plate recognition used in automatic vehicle identification systems (Shapiro et al., 2006; Anagnostopoulos et al., 2006; Abdullah et al., 2007), optical road estimation for autonomous vehicles (Dahlkamp et al., 2006), surveillance of traffic by video sensors (Kastrinaki et al., 2003), automatic detection of pedestrians and bicycles

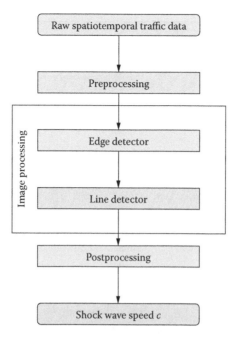

Figure 3.14 Structure of wave speed estimator. (*Source:* Schreiter, T., 2013, Vehicle-class Specific Control of Freeway Traffic, Ph.D. thesis, Delft University of Technology. With permission.)

(Malinovskiy et al., 2009) and vehicle detection (Chi and Caldas, 2011). In everyday life, image processing methods are used in face recognition applications in digital cameras.

Since the image processing methods operate on images, conversions from traffic data to image data and back are necessary. These are performed by the pre- and postprocessing components of the WSE. The remainder of this section explains each component of the WSE in detail. For illustration purposes, the intermediate results of each component are displayed in Figure 3.15.

Preprocessing — Before applying the image processing tools, the spatiotemporal traffic data are converted to an image. An image is a matrix in which each element represents a so-called pixel, which is a rectangular area of fixed size and is assigned one value representing brightness. In colored images, each pixel is assigned a vector of three values, representing red, green, and blue intensities, respectively. The image is created by interpolating the traffic data at equidistant sample points with a fixed resolution in space Δx_{discr} and time Δt_{discr}.

Furthermore, the raw traffic data may contain outliers and other high-frequency noises that obstruct the detection of the waves. A moving-average low-pass filter is therefore applied to reduce the high frequencies.

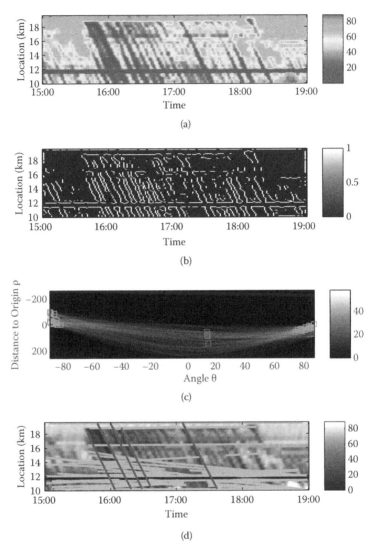

Figure 3.15 Intermediate results of wave speed estimator for shock wave speeds at stop-and-go waves. (*Source:* Schreiter, T., 2013, Vehicle-class Specific Control of Freeway Traffic, Ph.D. thesis, Delft University of Technology. With permission.)

Figure 3.15a shows the result of applying the preprocessing steps to the raw data from Figure 3.12a. Due to the low-pass filtering, the results look smoother than the original raw data, but the borders between high and low speeds are preserved so that shock waves are still visible.

Edge detector — The next step is to detect the transitions of traffic states. An edge detector can detect the transitions between dark and light regions

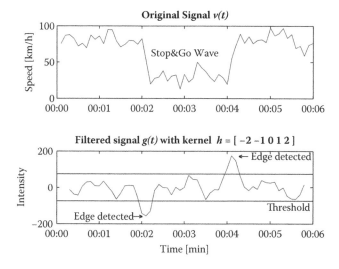

Figure 3.16 Example of edge detection in one dimension. (*Source:* Schreiter, T., 2013, Vehicle-class Specific Control of Freeway Traffic, Ph.D. thesis, Delft University of Technology. With permission.)

in an image. Figure 3.16 depicts the application of an edge detector in one dimension. The top diagram shows a raw traffic signal $v(t)$ over time, for example, from a stationary detector. Two traffic state transitions occur: at 00:02, traffic slows from 80 to 30 Km/h; two minutes later, it recovers. These transitions can be detected by a gradient-based method. Convolving the signal with a gradient-based kernel $h = [-2 - 1012]$ results in the signal

$$g(t) = (v \cdot h)(t) \tag{3.33}$$

shown at the bottom of Figure 3.16. The traffic state transitions are amplified and are visible by strong extremes. An edge detector applies the same principle in two dimensions. The gradients g_x and g_t are determined by applying a gradient-based kernel in each dimension, and then the gradients are superimposed:

$$G(x,t) = \sqrt{g_x(x,t)^2 + g_t(x,t)^2} \tag{3.34}$$

If the superimposed gradient G exceeds a specified threshold γ in a point (x,t), then an edge is present at the point:

$$E(x,t) = \begin{cases} 1 & \text{if } G(x,t) > \gamma \\ 0 & \text{else} \end{cases} \tag{3.35}$$

The result of an edge detector is thus a binary image E, representing the edges of the traffic state conditions. An example is shown in Figure 3.15b, where the diagonal lines represent the edges caused by the stop-and-go waves. Note that the absolute values of the original signal are irrelevant as long as the state transitions are visible. For example, the dual loop detectors used in the Netherlands are inherently biased because they aggregate speed arithmetically over one minute, which leads to overestimation (Knoop et al., 2009). By applying an edge detector, however, the state transitions can be located despite this bias.

Line detector — In the resulting image of the edge detector, the traffic regime transitions are visible as white lines. The ones caused by stop-and-go waves are straight lines. To locate them, a line detector is applied to the edge image.

The Hough transform (1962) is a widely used method of detecting lines. It transforms an original image from the Cartesian $x - t$ plane into the so-called Hough domain or $\rho-\theta$ plane. These parameters specify the polar coordinates of straight lines in the original picture by the relation

$$\rho = t\cos\theta + x\sin\theta \tag{3.36}$$

with the angle θ and the distance ρ to the origin of the image. Figure 3.17 illustrates the important properties of the Hough transform. A point (x,t) in the original picture corresponds to a sine wave in the Hough domain, as Figure 3.17a shows. Moreover, a set of collinear points corresponds to a set of sine waves that all intersect at exactly one point, as Figure 3.17b shows.

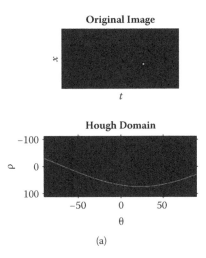

(a)

Figure 3.17 Relation between Cartesian $x - t$ image and its transform in Hough plane. (*Source*: Schreiter, T., 2013, Vehicle-class Specific Control of Freeway Traffic, Ph.D. thesis, Delft University of Technology. With permission.) *(continued)*

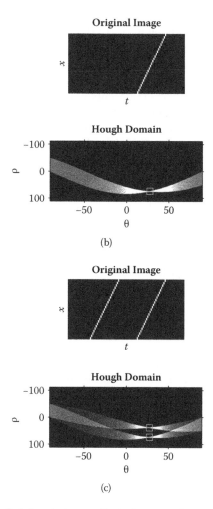

Figure 3.17 (continued) Relation between Cartesian *x – t* image and its transform in Hough plane. (*Source:* Schreiter, T., 2013, Vehicle-class Specific Control of Freeway Traffic, Ph.D. thesis, Delft University of Technology. With permission.)

This point (θ, ρ) defines the angle θ and the distance to the origin ρ of its corresponding straight line. Two parallel lines therefore result in a set of sine waves that intersect in two points with the same angular coordinate, as Figure 3.17c shows. The propagation speed of the corresponding shock waves is therefore determined by the angle θ:

$$c = -\tan(90° - \theta) \cdot \frac{\Delta x_{\text{discr}}}{\Delta t_{\text{discr}}} \tag{3.37}$$

The distance ρ of the line to the origin of the Hough plane (lower left corner) is used to locate the shock wave. In the case of standing waves, this can be used to locate the cause such as a recurrent bottleneck or accident. The Hough transform is applied to the output image of the edge detector. This leads to a grayscale image in the Hough domain, as shown in Figure 3.15c. The local maxima, which are the points with the highest intensity (boxes), are detected in the Hough domain. The results of the Hough transform and the line detector thus constitute a set of lines.

An example of the result of the line detector is shown in Figure 3.15d. The background is a grayscale version of the preprocessed image in Figure 3.15a.

Postprocessing — The result of the line detector is thus a set of lines. If the goal is to determine the shock wave speed that characterizes a specific traffic flow pattern, these lines must be analyzed further. Each traffic pattern is characterized by a prior probability density function (pdf) f^p that describes which wave speeds it causes. The lines found by the line detector then represent a second pdf.

$$f^L(c) = \frac{1}{\sum_{i=1}^n s_i} \cdot \sum_{i=1}^n s_i \cdot \delta(c - c_i) \tag{3.38}$$

with their corresponding propagation speeds $c_1, c_2, \ldots, c_n$ weighted by their corresponding lengths $s_1, s_2, \ldots, s_n$, and the functional δ as the Dirac pdf:

$$\delta(c) = \begin{cases} +\infty & \text{if } c = 0 \\ 0 & \text{else} \end{cases} \tag{3.39}$$

$$\int \delta(c) dc = 1 \tag{3.40}$$

By applying Bayes' law, the estimated posterior pdf:

$$f^e(c) = \frac{f^L(c) \cdot f^p(c)}{\int f^L(c) \cdot f^p(c) dc} \tag{3.41}$$

describes the distribution of filtered shock wave speeds. The mean of that pdf then is the mean shock wave speed:

$$c = \int f^e(c') \cdot c' dc' \tag{3.42}$$

which is the result of the WSE.

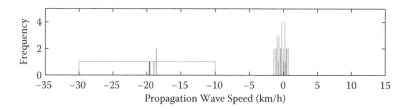

Figure 3.18 Frequency of wave speeds of lines detected; right: lines within prior distribution (gray) of Figure 3.15, left: all other lines detected. (*Source:* Schreiter, T., 2013, Vehicle-class Specific Control of Freeway Traffic, Ph.D. thesis, Delft University of Technology. With permission.)

In the ongoing example of Figure 3.15d, the shock waves of stop-and-go waves are of interest. Since stop-and-go waves propagate with a speed between −25 and −15 Km/h (Schönhof and Helbing, 2009; Kerner and Rehborn, 1997; Treiber et al., 2000; Kerner, 1998; Chiabaut and Leclercq, 2011; Bertini and Leal, 2005; Graves et al., 1998; Zhang and Rice, 1999; Cassidy and Mauch, 2001), we choose a uniform pdf:

$$f^p(c) = \begin{cases} \dfrac{1}{b-a} & \text{if } c \in [a,b] \\ 0 & \text{else} \end{cases} \tag{3.43}$$

in the range of $[a,b] = [-30 \text{ Km/h}, -10 \text{ Km/h}]$ as prior distribution f^p. Figure 3.18 shows a histogram of wave speeds of all lines detected (f^L). The gray function is defined by the prior pdf f^p. The posterior distribution f^e therefore contains all the lines that fall within the prior function. The mean of the posterior distribution is $c = -19.9$ Km/h.

3.4.2.3 Extracting shock waves from empirical data

In this section, the proposed method is validated for empirical data by applying it to spatiotemporal data gathered from Netherlands freeways. Three experiments were performed to estimate the shock waves caused by stop-and-go waves, by vehicle platoons in free flow, and by a fixed bottleneck. To evaluate the performance of the WSE, its estimated shock wave speeds were compared to those found by manual inspection. We generated the latter speeds by visualizing the raw data and estimating the shock wave speeds by hand. Finally, a sensitivity analysis was performed.

The WSE was applied to freeway data gathered from the southbound A13 between the Hague and Rotterdam (Km 10 to Km 20, Figure 3.19). The dual-loop detectors were placed approximately 500 m apart on average and aggregated the speed and flow data over one minute. The evening

Figure 3.19 Freeway A13 near Delft, Netherlands. (*Source:* Google Maps 2012.)

traffic data between 15:00 and 19:00 for 14 days (April 24 through May 7, 2009) served as raw data. Figure 3.12 shows two examples of the raw data. Since some of the days examined fell on weekends or holidays, not all traffic patterns occurred on all days.

3.4.2.4 Extracting shock waves from stop-and-go waves

Stop-and-go waves are common phenomena during congestion, as Figure 3.12 demonstrates. Speed data are used as input because stop-and-go waves are clearly visible there. The prior distribution f^p of the postprocessing component is set to a uniform function (3.29) with a range of $[a,b] = [-30$ Km/h, -10 Km/h].

Figure 3.20 presents the results of the WSE. For both wave speeds c_{cong} (this section) and c_{free} (Figure 3.20), the estimated wave speed (crosses) and the minimum and maximum speed values of all lines detected (triangles) are plotted. For comparison, the results of the manual inspection method are plotted as well.

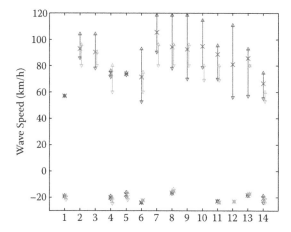

Figure 3.20 Results of wave speed estimator applied to empirical data to detect the speeds of shock waves (kilometers per hour). Crosses: average wave speed estimated; triangles: minimum and maximum values found; right: congestion (Figure 3.19); up: free flow (Figure 3.20); light gray results of manual inspection. (*Source:* Schreiter, T., 2013, Vehicle-class Specific Control of Freeway Traffic, Ph.D. thesis, Delft University of Technology. With permission.)

Congestion occurred on 9 of the 14 days. For 8 of the 9 days, the WSE estimated the congested wave speed c_{cong} within a range of 2 Km/h of the manual method. On the 12th day, no lines of congestion were detected. Since the values estimated by the WSE were close to the values of the manual inspection method, we conclude that the WSE is a good estimator of the speed of shock waves in congestion.

As the minimum and maximum values indicate, the lines detected were within a very small range; Figure 3.18 shows a histogram of all lines detected for the first data set. This narrow distribution of lines indicates that the shock waves occurring in congestion can be well summarized by one wave speed value c_{cong} for that day.

3.4.2.5 Extracting shock waves from free flow

Vehicle platoons and other fluctuations in free flow cause shock waves that propagate downstream. Since they are visible, spatiotemporal flow plots (see Figure 3.12) showing flow data are used as inputs. The prior distribution f^p is set to a uniform function (Equation (3.43)) with a range of $[a,b] = [50 \text{ Km/h}, 120 \text{ Km/h}]$. An example result of the WSE for the second day is shown in Figure 3.21a. Many shock waves are detected correctly. These shock waves propagate at different speeds, as the histogram in Figure 3.21b shows. Similarly, the other days show a large range of wave speeds as well (Figure 3.20). This result suggests that the free flow branch of the fundamental diagram is bent.

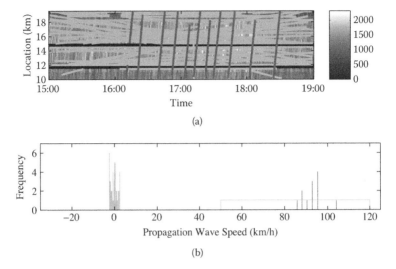

Figure 3.21 Results of extracting shock waves occurring in free flow. (*Source:* Schreiter, T., 2013, Vehicle-class Specific Control of Freeway Traffic, Ph.D. thesis, Delft University of Technology. With permission.)

In the case of free flow, the waves are difficult to detect manually. The results of the manual inspection are therefore very imprecise. Nevertheless, the results of Figure 3.20 show that the results of the WSE match the manual results in many cases. We conclude therefore that the WSE is also capable of estimating shock waves in free flow.

In this section, we showed a method to automatically detect shock waves and their propagation speeds from spatiotemporal traffic data. The method uses image processing to detect the transitions between traffic states. Since only the transitions are detected but not the absolute values of the traffic state, this method is robust toward biased data as the transitions occur in speed measurements of the freeways in the Netherlands. By averaging the propagation speed of the shock waves, the characteristic wave speeds of traffic regimes can be estimated.

One application of the characteristic shock wave speeds is the estimation of parts of the fundamental diagram. These are important parameters of traffic state estimators. The method can be used to automatically calibrate the ASM presented earlier in this chapter. Another application is the evaluation of the performances of traffic flow models. This is a recurrent task in model calibration. The WSE can thus be applied to calibrate traffic flow models intended to reproduce shock waves with the correct propagation speeds.

For further research, the number of stop-and-go waves occurring per time could be estimated if the WSE is fine tuned so that both borders of all stop-and-waves can be estimated. Furthermore, the composition of traffic could be estimated based on shock wave speed according to Equation (3.32). Research questions in this line concern the strength of the relationship between shock wave speeds and traffic composition observed in empirical data and how precisely the WSE must be able to detect the shock wave speeds to determine the traffic composition accurately and precisely.

3.4.3 Capacity estimation

As a final example of feature extraction, let us briefly consider capacity estimation. Estimating roadway capacity forms a key challenge for a number of reasons. First, the capacity is not a deterministic value; it is a random variable. Second, the capacity drop phenomenon causes the capacity to be dependent on the prevailing traffic conditions. In congested conditions, roadway capacity is 10 to 15% smaller than it is in free flow conditions. Third, the capacity observations are in many cases censored, which implies that we often do not observe free flow capacities. Instead, we observe flows that are by definition smaller than the prevailing capacity values.

In this section, we briefly discuss the capacity estimation problem and some of the approaches put forward to derive capacity values from empirical data.

3.4.3.1 Important estimation issues

Before discussing techniques to estimate roadway capacity, we propose some important issues that are relevant to assessing the capacity of a roadway or bottleneck.

Stochastic nature of capacity — Maximum flows (free flows or queue discharge rates) are not constant values and vary under the influence of several factors. Capacity is influenced, for example, by the composition of a vehicle fleet, the composition of traffic based on trip purpose, and weather, road, and ambient conditions. These factors affect the behavior of driver vehicle combinations and thus the maximum number of vehicles that can pass a cross section during a given time period. Some of these factors can be observed and their effects can be quantified. Some, however, cannot be observed directly.

Furthermore, differences among drivers mean that some drivers need more minimum time headway than other drivers even though they belong to the same user class. As a result, the minimum headway h_i values will not be constant but will follow a distribution function (see discussion of headway distribution modeling in Section 3.4.1).

Observed maximum flows thus appear to follow a distribution. The shape of this distribution depends on, among other issues, the capacity definition and measurement method and period. In most cases, a normal distribution may be used to describe capacity.

Capacity drop — Several researchers noted the existence of two different maximum flow rates, namely, pre-queue and queue discharge. Each rate has its own maximum flow distribution. We define the pre-queue maximum flow as the maximum flow rate observed at the downstream location just before the onset of a congestion queue upstream. These maximum flows are characterized by the absence of queues or congestion upstream of the bottleneck, high speeds, instability leading to congestion onset within a short period, and large variance.

The queue discharge flow is the maximum flow rate observed at the downstream location as long as congestion exists. These maximum flow rates are characterized by the presence of a queue upstream the bottleneck, lower speeds and densities, constant outflow, and a small variance that can sustain for a long period. However, flow rates are lower than in the pre-queue flow state.

Both capacities can be measured only downstream of the bottleneck location. Average capacity drop changes are in the range of –1 to –15%.

3.4.3.2 Approaches to capacity estimation

Many approaches may be applied to compute the capacity of a specific piece of infrastructure. The suitability of the approach depends on a number of factors such as:

1. Type of infrastructure (motorways with or without off and on ramps, roundabouts, unsignalized intersections, etc.)
2. Type of data (individual vehicle, aggregate) and time aggregation
3. Location of data collection (upstream of, within, or downstream of bottleneck)
4. Traffic conditions for which data are available (congestion, no congestion)

The two methods described below are available for estimating capacity. They do not require capacity observations.

Estimating composite headway models — The observed headway distribution is used to determine the distribution of the minimum headway, which in turn is used to estimate a single capacity value (no distinction between pre-queue capacity and queue-discharge rate). The main advantage is that no capacity observations are required.

Fitting a fundamental diagram — This approach uses the relationship between speed and density or flow rate to estimate capacity value. A functional form needs to be selected and assumptions about the critical density must be made.

The following methods require capacity observations.

Use of cumulative curves — When a bottleneck is oversaturated, the slope of the cumulative curve indicates the queue discharge rate. This approach is similar to the queue discharge distribution method.

Selected maxima — Measured flow rate maxima are used to estimate a capacity value or distribution. The capacity state must be reached during each maxima selection period. This approach should be applied over a long period.

Bimodal distribution — This method may be applied if the observed frequency distributions of the flow rates exhibit a clear bimodal form. The higher flow distribution is assumed to represent capacity flows.

Queue discharge distribution — This is a very straightforward method using queue discharge flow observations to construct a capacity distribution or capacity value. It requires additional observations (e.g., speeds upstream of the bottleneck) to determine the congested state.

Simulation — This approach is often used to determine the capacity of a roundabout or unsignalized intersection. It is based on combining the empirically determined critical gap distribution and headway distribution models.

Product limit — This method uses below-capacity flows together with capacity flows to determine a capacity value distribution. Speed and/or density data are needed to distinguish the type of flow measurement at the road section upstream of the bottleneck.

3.4.3.3 Queue discharge distribution method

The queue discharge distribution (or empirical distribution) method is based on the notion that the available measurements (downstream of a bottleneck) of flow can be divided into:

- Measurements representing traffic demand
- Measurements representing capacity state of the road
- Measurements representing capacity state of an active bottleneck upstream of the measurement site

In the latter case, the measurement cannot be used for capacity estimation. In practical studies, it is very important to identify these measurements. The queue discharge rate uses only data of the capacity state. These are usually identified by considering measurements upstream of the bottleneck and checking whether congestion occurs there. Figure 3.22 shows an application example using synthetic data.

Empirical distribution is a straightforward estimation method that yields an unbiased estimate of capacity distribution (or, more specifically, the queue discharge rate). The most important drawback is that the approach does not use any information that may be contained in non-capacity observations. Including this information can be achieved by applying the product limit method, as illustrated in the next section.

3.4.3.4 Application of product limit method (PLM) to capacity estimation

The rationale for using the PLM (or Kaplan–Meier approach) for capacity is the following: given that capacity is a random variable that changes

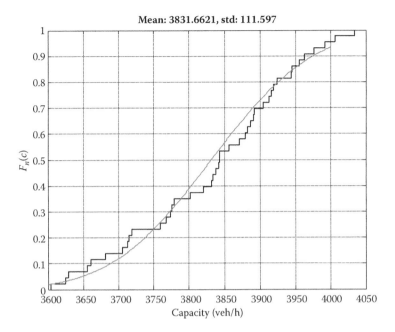

Figure 3.22 Example application of queue discharge distribution method using synthetic data. The synthetic queue discharge rates were drawn from normal distribution with a mean of 3800 vehicles per hour and standard deviation of 100 vehicles per hour. (*Source:* Hoogendoorn, S.P. and Van Lint, H. 2010. Contribution to CIE4831. Delft University of Technology. With permission.)

from one period to the next, traffic is more likely to break down when the capacity is small (capacity observation) than when it is high. This means in theory that the empirical distribution function of the queue discharge rates may be biased to low capacities. Furthermore, all information from the non-capacity observations is discarded.

We already pointed out that the PLM approach can also be used to estimate capacity. This is important, since it also uses measurements of non-capacity situations and can remove the bias to lower capacity observations. The nonparametric version is presented here and involves the following steps:

1. Determine a congestion criterion such that a capacity observation is defined by $u_i < u_{\text{cong}}$ (for example, $u_{\text{cong}} = 50, 70,$ or 90 km/h).
2. Split the data set into capacity observations ($u_i < u_{\text{cong}}$) and non-capacity observations ($u_i \geq u_{\text{cong}}$). The capacity observations are denoted by the set $\{C_i\}$ and the non-capacity observations by the set $\{Q_i\}$.

3. For each capacity observation C_i, let d_i = number of observations equal to C_i and m_i = sum of (1) number of capacity observations $\{C_i\}$ ≥ C_i and (2) number of non-capacity observations $\{Q_i\}$ > C_i.
4. Compute the survival function of the capacity $(S(c) = \Pr(C > c) = 1 - F(c))$:

$$S_{PLM}(c_i) = 1 - F_{PLM}(c_i) = \prod_{j=1}^{i} \left(\frac{m_j - d_j}{d_j} \right) \tag{3.44}$$

Note that the survival function is determined only at the capacity points C_i. We can easily show that if we have only capacity observations, the PLM estimate of the capacity and the empirical distribution method yield the same results.

Figure 3.23 shows the results of application of the PLM (parametric and nonparametric) for synthetic data generated with the Quast model. Clearly, the mean queue discharge rate (3800 vehicles/h) and standard deviation (100 vehicles/h) are much better approximated using PLM than using the empirical distribution method discussed in the previous section.

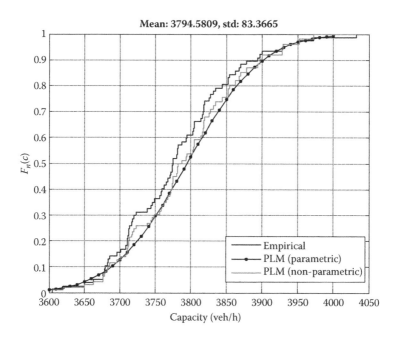

Figure 3.23 Estimation results of parametric and nonparametric application of PLM. (*Source:* Hoogendoorn, S.P. and Van Lint, H. 2010. Contribution to CIE4831. Delft University of Technology. With permission.)

3.4.3.5 Estimating capacities using fundamental diagram

In light of the capacity drop, two capacities exist: free flow and queue discharge. Both values are relevant and require different estimation methods.

Estimating free flow capacity — Capacity can be estimated by fitting a model of the function $q(k)$ to data. The calculated maximum of $q(k)$ is an estimate of capacity. This model is not suitable for modern motorway traffic where it does not hold that the model for the free flow part of $q(k)$ has a maximum with the derivative $dq/dk = 0$ at capacity.

An alternative method is to assume a value of the capacity density k_c, fit a curve to a model for the free flow part of $q(k)$ and estimate the capacity by $q(k_c)$. It is obvious that the outcome of the estimation depends on the value of k_c, and this makes the methods more suitable for comparisons than for absolute values. The procedure has been applied to determine the effect of road lighting on capacity.

Estimating queue discharge capacity — This capacity can be estimated only in an overloaded bottleneck. Ideally three measuring stations are required:

> Upstream of the bottleneck: data from this cross section can be used to determine whether overloading occurs (mostly mean speed is used as a criterion).
> Downstream of the bottleneck: the state of traffic flow should be free at this cross section. If this is not the case, a more severe bottleneck is present downstream.
> At the bottleneck: intensities are capacity measurements if the traffic state upstream is congested and the state downstream is free.

Remark 1 — At an overloaded bottleneck, intensity at all three measuring stations is the same. It is an advantage to use the intensity at the downstream section to estimate capacity because traffic flow is smoother than at the other stations and this reduces measuring errors.
Remark 2 — The fact that large parts of motorway networks all over the world exhibit daily overloading is helpful for analyzing capacity values.

3.5 ORIGIN–DESTINATION (OD) ESTIMATION

Transport planning has traditionally relied on static origin–destination (OD) matrices obtained via surveys or other aggregate methods. OD matrices suffer from limitations associated with their spatial coarseness and lack of time-dependent information. In terms of spatial network representation, the loading points of a network must be reduced significantly to accommodate the centroids considered by the OD matrix. Dealing with temporal limitations has been easier, typically using demand profiling according to available measurements. These approaches have been coarse and served as predecessors of the more elaborate OD estimation problem that uses traffic surveillance (typically link counts from loop detector data).

The problem of OD matrix estimation (and prediction) has been well studied during the last 30 years. A complete exploration of the relevant methods combining OD historical demand information and link count data found in literature is presented in this book ($ 6,2, pp 142–153). However a number of recent traffic surveillance technologies utilize different technical characteristics, including data collected, accuracy of the measurements, levels of maturity, feasibility and cost, and network coverage. These technologies transformed the applications of the OD estimation process. We will devote the next few paragraphs to a description of those new methods for OD estimation.

Ashok (1996) introduced direct measurements for the incorporation of probe vehicle information for solving the OD estimation problem. By definition, a direct measurement provides a preliminary estimate of an OD flow. In the case of information obtained from probe vehicles (with known origins and destinations), direct measurements may be represented by the product of the flow of probe vehicles with an appropriate expansion matrix to account for the fact that probe vehicles constitute only a fraction of the total vehicles in the network.

Van der Zijpp (1996) combined volume counts with trajectory information obtained from automated license plate surveys to estimate OD flows. A measurement equation for the trajectory counts is specified and split probabilities are estimated from combined link volume counts and trajectory counts. The method of estimating split probabilities from combined data was tested by the authors in experiments using synthesized data. These experiments showed that using combined data consistently led to lower errors relative to the case with only traffic counts.

Mishalani et al. (2003) evaluated the roles of various types of surveillance data in the real-time estimations of dynamic OD flows at the intersection level. Turning fraction and travel time data were collected from video cameras at three adjacent intersections. OD estimation using link counts served as the base case. Additional scenarios involved the incorporation of turning fractions at the intersections, travel times, and combinations of turning fractions and travel times. The main finding of the study was that intersection turning fraction data significantly improves the quality of the OD flow estimates, while the use of link travel time data in addition to turning fractions can also be useful.

Oh and Ritchie (2005) developed a methodology for the design of advanced traffic surveillance systems based on microscopic traffic simulation. The authors demonstrate the proposed methodology through the development of an inductive signature-based anonymous vehicle tracking system for the OD flow estimation problem.

Dixon and Rilett (2000) proposed a method for using sample link choice proportions and sample OD matrix information derived from automatic vehicle identification (AVI) data sampled from a representative vehicles to estimate population OD matrices with the AVI data collection points acting as the origins and destinations. This assumption, however, is limiting in terms of the applicability of the model. The authors also propose a method that extends their approach to the estimation of ramp-to-ramp freeway OD volumes.

Kwon and Varaiya (2005) developed a statistical OD estimation model using partially observed vehicle trajectories obtained with vehicle reidentification or AVI techniques such as electronic tags. The authors derived an unbiased estimator of the OD matrix based on the method of moments and developed a bootstrap standard error estimate of the OD matrix estimator.

Antoniou et al. (2004) presented a methodology for the incorporation of AVI information into the OD estimation and prediction framework, later generalized (Antoniou et al., 2006) into a flexible formulation that can incorporate a large range of additional surveillance information as it becomes available.

3.6 SUMMARY

This chapter focused on data processing and data enhancement techniques. Although not comprehensive, the chapter discussed various techniques for estimation and data fusion at different inference levels. At core level, the processing of the raw data was considered both for microscopic (trajectories) and macroscopic traffic flow data. At the second level, we reviewed examples showing how features of traffic flow can be derived from processed data. Since the third level that deals with decision making and event detection was beyond the scope of this book, we have not considered examples at that level in detail.

This chapter has shown which techniques are available for generating input for performing traffic simulation exercises. In doing so, we only considered approaches that to a large extent can make such inferences directly from processed data. In Chapters 4 and 5, we will go into more detail for determining model parameters by calibrating models to enhanced data.

Chapter 4

Calibration and validation principles

Christine Buisson, Winnie Daamen, Vincenzo Punzo, Peter Wagner, Marcello Montanino, and Biagio Ciuffo

CONTENTS

This chapter discusses the principles of calibration and validation and links these phases to the previous chapters on data collection and data enhancement. First, a generic outline of a typical simulation study is given in Section 4.1, then a generic procedure for calibration and validation is covered in Section 4.2, underlying three main choices: measures of performance, measures for goodness of fit, and optimization procedure. Section 4.3 presents examples of calibration for car following. Section 4.4 explains which role can be played by synthetic data while scrutinizing if the calibration and validation chain procedure is or not appropriate. Section 4.5 discusses conclusions.

4.1 TYPICAL STEPS OF SIMULATION STUDIES

Traffic simulation models are typically used for (1) assessment and planning of (road) infrastructures; (2) evaluation of advanced traffic management and information strategies; and (3) testing technologies and systems intended to increase safety, capacity, and environmental efficiency of vehicles and roads. A simulation study usually involves a comparison of a current and a future situation, with possible modifications in the network (infrastructure), traffic management, information strategies, and relevant technologies or systems.

In addition, the effects of external conditions such as increased demand or changes in traffic composition are investigated. The simulation task involves multiple steps, depending on the specific study questions. Most authors' descriptions of the successive steps of a typical calibration study converge (see Dowling et al., 2004, for an example). A typical task list is:

1. Define the objectives of the study and the alternative scenarios to be tested.
2. Define the measures of performance that will be used to compare the current situation with the alternative.
3. Define the network to simulate by:
   a. Characterizing links: number of lanes, lengths, desired speeds, and upstream and downstream nodes.
   b. Characterizing nodes: allowed turning movements, upstream and downstream links, signalization (traffic lights with fixed phases, adaptive controllers, roundabouts, priority rules).
4. Define demand; this can be done from an additional model (usually static) or through the use of the existing data (in this case, measured flows on several links are needed) or by a combination of both.
5. Run the simulation and check whether the model performs as expected (verification).
6. Collect the data for calibration and validation by:
   a. Collecting the data set that allows definition of simulation entry variables (both static and dynamic).
   b. Collect the measures of performance that will allow comparison of simulation results with the observed current reality.
7. Calibrate and validate the traffic simulation tool for the specific site and the reference scenario.

8. Simulate the alternative scenarios; based on specific cases, describe at least one of the following: (a) new infrastructure; (2) new regulations; (3) new demand.
9. Analyze the impact of the scenario on the simulation results by carefully scrutinizing the impact of the evolution of the scenarios on the chosen measure of performance.
10. Write the report.

During a survey performed within this MULTITUDE project (2013), we evaluated whether calibration and validation steps are performed by users and received more than 300 responses worldwide. More than half of the respondents did not perform calibration. A few pretended to validate without calibrating. However, using a simulation tool without calibration and validation raises questions on the predictive capacity of the tool.

When building the reference and alternative scenarios, one must define which traffic phenomena or behaviors should be included in the study and should be accurately described by the simulation model. For example, a tool does not necessarily have to describe the queue formations upstream of roundabouts if the researcher wants to evaluate the impact of ramp metering installations on a highway, but an accurate lane changing process is essential. A disaggregate data analysis is necessary for each key behavior included in a simulation scenario. Optimally, this detailed data analysis is performed and reported by the developers of the simulation tool. The user can thus rely on this disaggregate testing of each component to establish the suitability of the tool for use in the specific scenario planned.

Along with a detailed evaluation of the predictive capacity of the tool for each key behavior implied in the scenario to be simulated, aggregate calibration and validation must be completed for the application scenario. The calibration and validation of the model should focus on the specific site and traffic situations to be covered in the simulation study. Depending on the site chosen for the simulation, several variables should be considered:

- Type of network: size (number of links and nodes); urban or interurban routing, with and without traffic signals, roundabouts, curves, ramps, and combinations,
- Conditions of use: morning and evening peak hours, weekends, and holidays; weather conditions, traffic composition (percentages of trucks and passenger vehicles), evacuation needs,
- Traffic management system: adaptive control for traffic signals, driver information collected by means of GPS or by on-board devices, for individual information anti-collision, and other advanced driver assistance systems (ADAS).

With respect to traffic management systems, this component of calibration and validation focuses more on the settings of the systems since the disaggregate calibration and validation already demonstrated that the simulation model can reproduce these technologies and systems in general.

The next section introduces a complete calibration and validation framework. It combines data collection, identification of the measures of performance and goodness of fit, and the relationship of calibration and validation.

4.2 GENERIC PROCEDURE FOR CALIBRATION AND VALIDATION

Globally, calibration and validation of a simulation tool (with a set of parameters) is a process of comparison on an appropriate scale of the simulation results for chosen variables with a set of observations of the variables. (See Figure 4.1.) During calibration, the differences between the simulated and observed variables are minimized by finding an optimum set of model parameters. This is often formulated as minimizing an error measure $e(p)$ (Vaze et al., 2009; Ciuffo and Punzo, 2010a):

$$e(p) = f(M(p) - d) \tag{4.1}$$

d is the observed measurement.

where the model M can, of course, depend on much more than parameters p. Thus, calibration becomes an optimization problem. The validation process estimates the difference between the simulation variables using the parameter set, resulting from calibration and an independent set of observed variables. In the following, a variable is referred to as a measure of performance (MoP). The comparison scale is the goodness-of-fit measure (GoF). The practical specification of calibration and validation relies on several factors:

- Error function, generally GoF
- Traffic measurement, usually called the MoP of the simulation model
- Optimization method for calibration
- Traffic simulation model used
- Transportation system (traffic scenario) to be simulated
- Demand pattern used and accuracy of its estimation
- Parameters to calibrate
- Quality (in terms of error presence) of observed measurement
- Possible constraints

The existence of these factors makes calibration and validation more complicated because it is difficult to define the general methodologies to be followed in all the cases. Depending on the particular specification of the

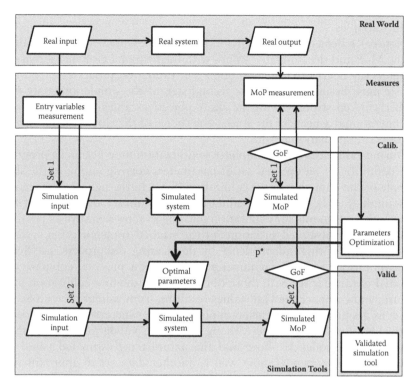

Figure 4.1 Depiction of global framework of calibration and validation. The first data set (Set 1) is used for the calibration procedure. The successives steps of the parameter optimization procedure are represented with bold arrows. The second subset is used for validation (Set 2). The set of optimal parameters ($\hat{p}$), resulting from calibration serves as model parameters during the validation. When comparing the measured and simulated measure of performance (MoP) during the validation process, one must determine the distance between them measured with the help of the same goodness of fit (GoF) of the same amplitude at the end of the calibration process. (Source: Ciuffo, B. et al., 2012. The Calibration of Traffic Simulation Models: Report on the Assessment of Different Goodness-of-Fit Measures and Optimization Algorithms. MULTITUDE Project and JRC Scientific and Technical Reports. With permission.)

calibration or validation, the strategy to be adopted in practice may be different. However, the calibration and validation framework shown in Figure 4.1 can still be maintained.

For a specific traffic scenario, after defining the transportation network and corresponding demand pattern, the sensitive parameters of the model can be identified by means of a sensitivity analysis as shown in Chapter 5.

Then, decisions should be made on the MoP, GoF, and optimization algorithm variables.

Figure 4.1 depicts the comparison between measurement of real values of the MoP and the simulated values with the help of a GoF measurement, which is the heart of both calibration and validation processes. The MoP choice must be made during the second step of the simulation study (see earlier), linked strongly with the study objectives and in agreement with the operational aspects of the study. The choice of the GoF is technical and impacts the simulation results.

During calibration, the parameter optimization loop permits a progressive definition of an optimal set of parameters corresponding to the subsample of the data chosen for calibration (Set 1 in the figure). The second subsample (Set 2) is used for validation. To minimize the impact of the choice, the subsample for calibration must be as representative as possible of the various observed situations of the studied transportation system. One can ensure this independence by duplicating the process: use Set 2 for calibration, reset the parameter optimization process, compare the optimal parameter set with those obtained from the first calibration procedure, and compare the GoF values resulting from validation with Set 1.

Let us designate the parameter set obtained after successful calibration as $\hat{p}$. Calibration is incomplete without validation. Validation asks whether, based on the model M above and the parameters $\hat{p}$ estimated based on observed measurement d, how well is another data set d' approximated by the model? To answer this, once more $e(p)$ must be computed, but now without the minimization of $e(p)$ done for calibration.

$$e_v = e(\cdot, \hat{p}) \tag{4.2}$$

In conclusion, the limited capacity of our models to depict reality makes it necessary, for their correct use, to follow a long iterative process of continuous verification of the steps, as in a Hegelian process in which a concept or its realization passes over into and is preserved and fulfilled by its opposite. The remainder of this chapter will provide more detail about the various steps of the calibration and validation framework shown in Figure 4.1.

4.2.1   Defining measure of performance (MoP)

The MoP must be defined in strong interaction with the application of the simulation tool in mind and, most important, its objective. For example, if the objective of the network modification between reference and future scenarios is to improve the mode share, one has to use the mode percentage as the MoP instead of, for example, vehicle queue length. Also, the MoP must be observable in reality with the available measurement devices and must be easily calculable from the simulation outputs.

4.2.2 Defining and collecting data set

After choosing an MoP adapted to the objective of the work, the next steps are designing the experiment and collecting the data. We can distinguish between single-valued MoP (characterizing the global behavior of the simulation scenario with a single value) and multivalued MoP reflecting the evolution of the system during the simulation duration. When the latter MoP is chosen, care must be taken in defining aggregation periods: overly long aggregation periods may average out specific characteristics of the traffic system; aggregation periods that are too short may include too much noise.

The question of the coherence of the definitions of the observed and simulated MoPs should be carefully addressed, because the definitions may be the sources of multiple errors. A classic example is when an observed variable is defined as the arithmetic temporal mean speed and the output of the simulation is computed as the harmonic temporal mean speed (see Chapter 3 for an explanation of the difference).

Specifically, if extreme single-valued MoPs are used (such as maximum waiting time), the question of the sample representativeness is crucial and the MoP comparison must be made with an appropriate number of simulation runs. Table 4.1 proposes a list of MoPs and appropriate data collection procedures along with descriptions of their drawbacks and advantages. For an explanation of the data collection procedure, see Chapter 2. For techniques to filter and enhance data, see Chapter 3.

4.2.3 Measures of performance (MoPs)

4.2.3.1 Typical MoPs used in the literature

The measure of performance of a system can be defined as a collection of variables necessary to describe the status of the system (Law, 2007). Depending on a system's complexity, several MoPs may exist. However, since a system is usually observed for a specific purpose, the MoPs that best describe its status depend also on the analyses to be carried out. This concept also applies to transportation systems. Their specific characteristics thus influence the calibration and validation of a traffic simulation model.

Common MoPs for the calibration and validation of a traffic simulation model are time series of speeds and counts collected on a road section (possibly differentiated per lane) aggregated over a certain time interval (Hourdakis et al., 2003; Toledo et al., 2003; Kim and Rilett, 2003; Toledo et al., 2004; Chu et al., 2004; Dowling et al., 2004; Kim and Rilett, 2004; Brockfeld et al., 2005; Schultz and Rilett, 2005; Balakhrishna et al., 2007; Ma et al., 2007; Ciuffo et al., 2008; Lee and Ozbay, 2008; Menneni et al., 2008; Vaze et al., 2009; Punzo and Ciuffo, 2009; Ciuffo and Punzo, 2010b).

The other fundamental variable of the traffic, namely, density, is used less frequently for calibration and validation purposes (Dowling et al., 2004;

Table 4.1 Data collection techniques associated with measures of performance (MoPs)

MoP type	Name	Data collection device	Comment
	Flow	Any detector described in Section 2.2.2; typically a loop detector	Error can be as high as 20% on total flow; take care with aggregation period definition; if aim is to validate network-wide simulation (OD estimation), link coverage must be defined carefully
	Density	Any detector described in Section 2.2.2; typically a loop detector	Low-quality measurements; take care with aggregation period definition
Multivalued collective variables	Speed	Any detector described in Section 2.2.2; typically a loop detector	Low-quality measurements, especially for low speeds; take care with aggregation period definition
	Queue length	Visual observation or set of loop detectors with speed observations	Human errors generally reduce data accuracy with visual observation; for loops, the quality of queue length measurement relates directly to spatial densities of detectors
	Travel times	Floating car data (FCD) Mean loop measured speeds	With FCD, size and representativeness of the sample must be verified carefully; quality of travel times data relates directly to spatial densities of loops
Multivalued individual variables	Trajectories	Camera placed at elevated position with automatic numeration	Numerical treatment can lead to errors that must be corrected
	Individual travel times	FCD or individual vehicle identification	Representativeness of observation sample (coverage, duration, number of days) is crucial

(continued)

Table 4.1 Data collection techniques associated with measures of performance (MoPs) (Continued)

MoP type	Name	Data collection device	Comment
	Last decile waiting time	FCD or mean speed	Representativeness of observation sample (duration, number of days) is crucial
Single-valued	Last decile queue length	Manual observation or automatic loop detector	Human errors generally reduce data accuracy with visual observation; representativeness of observation sample (duration, number of days) is crucial
	Mode share	Manual observation	

Ma et al., 2007) because it is more difficult to observe. Usually, density data are derived from the occupancy data from a single detector, which is not really accurate.

Other measures that are frequently used on more specific studies are queue lengths and turning flows at intersections (Ma and Abdulhai, 2002; Park and Schneeberger, 2003; Toledo et al., 2003; Dowling et al., 2004; Merritt, 2004; Shaaban and Radwan, 2005; Oketch and Carrick, 2005). Finally, more recently, based on the availability of more detailed information, point-to-point network travel times have been studied, both as aggregated measures and as distributions (Park and Schneeberger, 2003; Toledo et al., 2003; Chu et al., 2004; Dowling et al., 2004; Kim et al., 2005; Park and Qi, 2005; Oketch and Carrick, 2005; Hollander and Liu, 2008b; Vaze et al., 2009).

Vehicle trajectory data may also be used for the calibration of traffic simulation models. Because trajectory data are difficult to collect, they are rarely used. However, since data from the NGSIM project (NGSIM, 2011) have been made available, new applications are possible (Chiu et al., 2010).

4.2.3.2 Criteria for MoPs selection

The criteria discussed below are useful for selecting MoPs for calibration and validation.

Context of the application — MoP statistics should be important in the intended study. For example, point-to-point travel times are useful MoPs for validation when a traveler information system is to be evaluated on the basis of travel time savings. However, if a sensor-based incident detection system is studied, MoPs extracted from sensors (occupancies, flows, speeds) may be more useful.

Independence — MoPs used for validation should be independent of any measurements used for calibration or estimating inputs to a simulated system. Origin–destination (OD) flows are commonly estimated by minimizing a measure of the discrepancy between observed and simulated traffic counts. Therefore, validation of the simulation model (only) against traffic counts may lead to overestimating the realism of the model.

Error sources — In traffic analysis, a discrepancy between observed and simulated outputs can be explained by the following sources of error (Doan et al., 1999):

- Travel demands (OD flows)
- Route choices
- Driving behaviors
- Measurement errors in observed outputs

The first three sources contribute to errors in the simulated output. The last source represents errors in the observed output relative to the true output. In most cases, the contributions of the three simulation error sources are confounded and cannot be isolated in a validation. The locations and types of MoPs to be collected should be chosen to reflect errors from all these sources and reduce the effects of measurement errors as much as possible. Measurement locations should provide spatial coverage of all parts of a network.

Moreover, measurements near the network entry points will usually reveal errors in the OD flows with little effects from route choice and driving behavior models. As many measurement points as possible should be used to reduce the effects of measurement errors, assuming that the measurement errors are independent for different locations.

Traffic dynamics — MoPs and the level of temporal aggregation at which they are calculated (15 minutes, 30 minutes) should be chosen to facilitate testing whether or not the model correctly captures the traffic dynamics. This is especially true in network applications where both the temporal and spatial aspects of traffic are important.

Level of effort required for data collection — In practice, this is often the most constraining factor. Point measurements (flows, speeds, and occupancies) are often readily available from an existing surveillance system. Other types of measurements (travel times, queue lengths, and delays) are more expensive to collect. It is also important to note that data definitions and processing are not standardized. For example, statistics such as queue lengths may be defined in various ways and surveillance systems may apply a number of time-smoothing techniques. It is therefore necessary to ensure that the simulated data are defined and processed the same way as the observed data.

Number of runs — Most traffic simulation models are stochastic (Monte Carlo) simulations. Hence, MoPs should be calculated from a number of independent replications. The two main approaches to determine the number of replications are sequential and two-step (Alexopoulos and Seila, 1998). In the sequential approach, one replication at a time is run until a suitable stopping criterion is met. Assuming that the outputs Y_i from different simulation runs are normally distributed, Fishman (1978) suggested the following criterion:

$$R \geq R_i = \max \left(2, \left(\frac{s_R(Y_i)t_{\alpha/2}}{d_i} \right)^2 \right) \tag{4.3}$$

Here, R is the number of replications performed and R_i represents the minimum number of replications required to estimate the mean of Y_i with tolerance d_i. $s_R(Y_i)$ is the sample standard deviation of Y_i based on R replications and $t_{\alpha/2}$ is the critical value of the t distribution at significance level α.

In the two-step approach, first an estimate of the standard deviation of Y_i is obtained by performing R_0 replications. Assuming that this estimate does not change significantly as the number of replications increases, the minimum number of replications required to achieve the allowable error d_i is given by:

$$R_i = \left(\frac{s_{R_0}(Y_i)\,t_{\alpha/2}}{d_i} \right)^2 \tag{4.4}$$

The required number of replications is calculated for all measures of performance of interest. The most critical (highest) value of R_i determines the number of replications required.

4.2.3.3 Choice of appropriate statistical tests for comparing simulated and observed MoPs

The general simulation literature includes several approaches for the statistical validation of simulation models. These approaches include goodness-of-fit measures, confidence intervals, and statistical tests of the underlying distributions and processes. In many cases, however, they may not be applicable because both the real and the simulated traffic processes of interest are nonstationary and autocorrelated. The choices of the appropriate methods and their application to the validation of traffic simulation models depend on the nature of the output data. The following methods and their outputs are considered:

- Single-valued MoPs (e.g., average delay, total throughput)
- Multivariate MoPs (e.g., time-dependent flow or speed measurements at different locations, travel times on different sections)

Single-valued MoPs are appropriate for small-scale applications in which one statistic may summarize the performance of a system. Multivariate MoPs capture the temporal and/or spatial distribution of traffic characteristics and thus are useful to describe the dynamics at the network level. It may also be useful to examine the joint distribution of two MoPs (e.g., flows and travel times) to gain more information regarding the interrelationships of MoPs. The next section describes a statistical tests needed to calibrate and validate a simulation tool.

4.2.4 Goodness of Fit (GoF)

A number of goodness-of-fit measures can be used to evaluate the overall performance of a simulation model. This section reports on the studies of Hollander and Liu (2008a) and Ciuffo and Punzo (2010a). Note that GoF methods can be used both for calibration and validation, and are reported here for both applications. Table 4.2 lists GoF measures used for the calibration of traffic flow models and indicates the works in which they have been used.

Ciuffo and Punzo (2010a) analyzed 16 GoF measures, in particular, using response surface techniques and their suitability for use as error functions in Equation (4.1) were investigated. The authors derived the following findings:

- Response surfaces confirm, as argued in the introduction, the complexity of the calibration problem and the need to use global optimization. Most of the response surfaces showed several local minima as well as wide areas with approximately constant values. Clearly, different choices in setting up a calibration problem generate different response surfaces.
- U_m, U_s, $-U_c$, and $-r$ (correlation coefficient) proved less suitable than other factors to be used in the objective function of the calibration. In particular, U_s, U_c, and $-r$ were always more irregular, showing several different minima in all plots.
- The values of 3 and 5 as thresholds in $GEH3$ and $GEH5$ evaluations proved to be very high, and, consequently, a wide area in all their plots generated a constant value of the objective function. This suggests that, at least in the transportation field, 5 is probably a too high threshold to assess whether two series of data show a good fit, as proposed by the Highway Agency (1996).
- All other GoFs showed similar behaviors on the whole, even if SE appeared to be the least sensitive GoF (widest deep area around the "true" solution) but also the most regular around the minimum value.
- $RMSE$ seemed to offer higher irregularity than the other GoFs around the optimum value.
- $GEH1$ probably showed the best capacity in highlighting the position of the minimum (in this case the threshold used seems to have a good impact).

Table 4.2 Measures of goodness of fit

Name	Measure	Comments
Percent error *(PE)*	$\dfrac{x_i - y_i}{y_i}$	Applied to single pair of observed and simulated measurements or to aggregate network-wide measures
Squared error *(SE)*	$\dfrac{1}{N}\sum_{i=1}^{N}(x_i - y_i)^2$	Most used GoF; low values show good fit; strongly penalizes large errors; serves as basis of least squares method which, according to Gauss-Markov theorem (Plackett, 1950), provides best parameter estimation for linear models with zero-mean, unbiased and uncorrelated errors
Mean error *(ME)*	$\dfrac{1}{N}\sum_{i=1}^{N}(x_i - y_i)$	Indicates systematic bias; useful when applied separately to measurements at each location; cannot be used in calibration because low values prevent good fit (same high errors with opposite sign will result in zero *ME*)
Mean normalized error *(MNE)* or mean percent error *(MPE)*	$\dfrac{1}{N}\sum_{i=1}^{N}\dfrac{x_i - y_i}{y_i}$	Indicates systematic bias; useful when applied separately to measurements at each location; cannot be used in calibration because low values prevent good fit (same high errors with opposite sign will result in zero *ME*)
Mean absolute error *(MAE)*	$\dfrac{1}{N}\sum_{i=1}^{N}|x_i - y_i|$	Not particularly sensitive to large errors
Mean absolute normalized error *(MANE)* or mean absolute error ratio *(MAER)*	$\dfrac{1}{N}\sum_{i=1}^{N}\dfrac{|x_i - y_i|}{y_i}$	Not particularly sensitive to large errors; using absolute values would result in using same weights for all errors (it is preferable to assign more importance to high errors); gradient of absolute value analytical function has discontinuity point in zero; second most used GoF
Exponential mean absolute normalized error *(EMANE)*	$A \exp(-B\ MANE(x, y))$	Used as fitness function in genetic algorithm; *A, B* are parameters
Root mean squared error *(RMSE)*	$\sqrt{\dfrac{1}{N}\sum_{i=1}^{N}(x_i - y_i)^2}$	Large errors heavily penalized; may appear as mean squared errors without root sign

(continued)

Table 4.2 Measures of goodness of fit (Continued)

Name	Measure	Comments
Root mean squared normalized error (*RMSNE*) or root mean squared percent error (*RMSPE*)	$$\sqrt{\frac{1}{N}\sum_{i=1}^{N}\left(\frac{x_i - y_i}{y_i}\right)^2}$$	Large errors heavily penalized; normalized measures (also *MANE*) are attractive GoFs since they allow model calibration using different measures of performance (only relative error is considered); instabilities due to low values among measurements in fraction denominator may affect their use
GEH statistic	$$\sqrt{2\frac{(x_i - y_i)^2}{x_i + y_i}}$$	Applied to single pair of observed and simulated measurements, not over data series; *GEH* < 5 indicates good fit; looks suspiciously like one term of χ^2 sum; when applied to time series, *GEH* < 5 for 75% of observed and simulated measurements indicates good fit (Highway Agency, 1996); *GEH* can also be used by counting the number of times its value is under a certain threshold; this avoids taking very small errors into account; threshold considered also defines error neglected); to remove dependence on number of available observations, number of times *GEH* is under threshold can be divided by total number of observations; in this way, indicator would be bounded between 1 (perfect fit) and 0 (worst fit)
Correlation coefficient (*r*)	$$\frac{1}{N-1}\sum_{i=1}^{N}\frac{(x_i - \bar{x})(y_i - \bar{y})}{\sigma_x \sigma_y}$$	
Theil's bias proportion (U_m)	$$\frac{N(\bar{y} - \bar{x})^2}{\sum_{i=1}^{N}(y_i - x_i)^2}$$	High value implies systematic bias; U_m = 0 indicates perfect fit; U_m = 1 indicates worst fit
Theil's variance proportion (U_s)	$$\frac{N(\sigma_y - \sigma_x)^2}{\sum_{i=1}^{N}(y_i - x_i)^2}$$	High value implies that distribution of simulated measurements is significantly different from that of observed data; U_s = 0 indicates perfect fit; U_s = 1 indicates worst fit
Theil's covariance proportion (U_c)	$$\frac{2N(1-r)\sigma_x \sigma_y}{\sum_{i=1}^{N}(y_i - x_i)^2}$$	Low value implies unsystematic error; U_c = 1 indicates perfect fit; U_c = 0 indicates worst fit; *r* is correlation coefficient

(continued)

Table 4.2 Measures of goodness of fit (Continued)

Name	Measure	Comments
Theil's inequality coefficient (U)	$$\frac{\sqrt{\frac{1}{N}\sum_{i=1}^{N}(y_i - x_i)^2}}{\sqrt{\frac{1}{N}\sum_{i=1}^{N}(y_i)^2} + \sqrt{\frac{1}{N}\sum_{i=1}^{N}(x_i)^2}}$$	Combines effects of all Theil's error proportions (U_m, U_s, U_c); $U = 0$ indicates perfect fit; $U = 1$ indicates worst fit
Kolmogorov-Smirnov test	$\max(|F_x - F_y|)$	F is cumulative probability density function of x or y; requires more detailed traffic measurements; when comparing time series, two series may show same distribution of values and thus good KS-test result, but may be completely different
Speed flow graph	$Y - (Y \cap X)$	Parameter combination allows simulated and observed speed flow diagrams to overlap
Moses and Wilcoxon tests		Detailed procedure described by Kim et al. (2005)

x_i = simulated measurements.

y_i = observed measurements.

N = number of measurements.

$\overline{x}, \overline{y}$ = sample average.

σ_x, σ_y = sample standard deviation.

X,Y area of speed flow diagram covered by simulated and observed measurements.

Sources: Adapted from Hollander, Y. and Liu, R. 2008a. *Transportation Research Part C*, 16(2), 212–231; Ciuffo, B. and Punzo, V. 2010a. Proceedings of 84th Annual Meeting. Washington, D.C.: Transportation Research Board. With permission.

- *MAE* and *MANE* show the highest flatness of the objective function around the global solution. This could represent a problem identifying an effective stopping rule for any optimization algorithm.
- All these GoFs proved robust with respect to the introduction of noise to the data even with large errors.

4.2.5 Optimization Algorithms for Calibration Procedure

To perform calibration, we apply an optimization process after choosing the parameters, the MoP, and the GoF. Equation (4.1) demonstrates a well-defined optimization problem. As shown in Figure 4.1, it is now necessary to choose an optimization algorithm capable of solving the problem. The choice is not straightforward, since the nature of the problem suggests the best algorithm to be used. In general the choice should be oriented toward an algorithm for global "black-box" optimization, but this does not simplify the selection. Possible candidates are:

- Simultaneous perturbation stochastic approximation (SPSA)
- Simulated annealing (SA)
- Genetic algorithm (GA)
- OptQuest/Multistart algorithm
- Nelder-Mead
- Evolutionary algorithms
- Gradient-based methods
- Second-order models
- Scatter search heuristic

In the following we outline the main features of some optimization algorithms. The reader interested in further details should consult the references cited throughout the sections.

4.2.5.1 Simultaneous perturbation stochastic approximation (SPSA_I and SPSA_II) algorithms

One optimization method that has attracted considerable international attention is the SPSA method developed by James Spall (1992, 1998, 2000, 2003, and 2009). It is a stochastic analogue of the Newton-Raphson algorithm of deterministic nonlinear programming and basically extends the stochastic approximation (SA) method of Kiefer and Wolfowitz (1952) used to find extremes of functions in the presence of noisy measurements.

The SA method numerically evaluates the gradient of a function using a standard finite difference gradient approximation (each step of the algorithm thus requiring $2\,p$ function evaluations where p is the number of variables of the function). SPSA improves the SA method since the gradient evaluation is based on only two measurements of the objective function (at each step of the procedure) in which all the variables of the function are simultaneously perturbed (and not one at a time as in SA).

As a consequence, SPSA has two features particularly relevant in the calibration of micro-simulation traffic models: (1) it explicitly takes into account the presence of measurement errors in the objective function and (2) its results are usually less expensive to obtain than results from many other algorithms. In addition, in the past decade, the performances of the algorithm have been mathematically demonstrated in the form of an extensive convergence theory, both for local optimization (Spall, 1992) and global optimization (Maryak and Chin, 2008).

This latter case is particularly interesting in the case of the calibration of a microscopic traffic simulation model. Ciuffo and Punzo (2010b) demonstrated that this calibration problem must be seen in the framework of global optimization. More details can be found in Maryak and Chin (2008) who demonstrated the strict convergence of the SPSA algorithm using noise injection and a convergence in probability without the application of this method. This means that SPSA may also be suitable also in global optimization, but this should be tested for each specific case under evaluation.

SPSA has been used for the calibration of traffic simulation models by Balakrishna et al. (2007), Ma et al. (2007), Lee and Ozbay (2009), Vaze et al. (2009), and Ciuffo and Punzo (2010a). The source code is accessible (Spall, 2001).

4.2.5.2 Simulated annealing (SA)

Simulated annealing is a well-known method for solving unconstrained and bound-constrained global optimization problems. It is based on the principle of annealing, meaning that the magnitudes of random perturbations (injected within the parameter domain in a Monte Carlo fashion) are reduced in a controlled manner. This method is designed to increase the probability of avoiding local minima in the path toward the global minimum of the function. The randomness injection (as already pointed out for the SPSA) helps to prevent premature convergence, by providing a greater "jumpiness" to the algorithm (Spall, 2003). The *annealing* term is an analogy to the controlled cooling of a physical substance to achieve a type of optimal state (lower energy configuration) of the substance itself.

The theoretical basis of the simulated annealing algorithm can be found in Kirkpatrick et al. (1983) and Spall (2003). SA has been used in several transportation applications (Chang et al., 2002; Chen et al., 2005; Ciuffo and Punzo, 2010a). MATLAB® code for the simulated annealing algorithm can be found in Vandekerckhove (2010) to implement the simulated annealing algorithm described by Kirkpatrick et al. (1983).

4.2.5.3 Genetic algorithm (GA)

GAs are probably the most famous evolutionary algorithms and the most widely used for calibrating microscopic traffic simulation models. GA use is straightforward since no information about the objective function is required for its application (suitable for black-box optimization). GA is a stochastic global search method for solving both constrained and unconstrained optimization problems. It is based on natural selection, the process that drives biological evolution. GAs continuously modify a population of individual solutions. At each step, the GA selects individuals at random from the current population to be parents and uses them to produce children for the next generation.

Over successive generations, the population evolves toward an optimal solution. A fundamental difference between GAs and the other algorithms used so far is that they work with a population of potential solutions to a problem. This increases the probability for the algorithm to find a global solution.

The scientific literature on genetic optimization is extensive so the fundamentals will not be repeated here. For a review of the topic, refer to Spall (2003), Hollander (1975), and Harik et al. (1999). GAs have been applied many times for the calibration of microscopic traffic simulation models; see Ma and Abdulhai (2002), Schultz and Rilett (2004), Kim et al. (2005), Ma et al. (2007), and Ciuffo and Punzo (2010a). The available source code is the Genetic Algorithm Toolbox (for use with MATLAB) developed by the University of Sheffield (Chipperfield et al., 2010).

4.2.5.4 OptQuest/Multistart algorithm

The OptQuest/Multistart heuristic (Ugray et al., 2005) is an optimization algorithm for solving both constrained and unconstrained global optimization problems. It is at the same time a scatter search heuristic and a gradient-based algorithm. It has been employed recently for the calibration of microscopic traffic simulation models in Ciuffo et al. (2008) and Ciuffo and Punzo (2010a).

Basically, the algorithm employs a scatter search metaheuristic (Glover, 1998) to provide starting points for a generalized reduced gradient NLP solver (Drud, 1994). In this way, it tends to combine the seeking behavior of gradient-based local NLP solvers with the global optimization abilities of a scatter search. In practice, the scatter search performs a preliminary exploration in the parameter domain to locate different starting points for the gradient descent (which converges to the "nearest" local solution).

Adopting a sufficient number of starting points yields a high probability of finding a global solution of an optimization problem. The major shortcoming of this approach is the large number of objective function evaluations (i.e., traffic simulations) usually required, particularly as the number of parameters to be calibrated increases. The OptQuest/Multistart algorithm is implemented in Lindo API (Lindo Systems, 2002).

4.3 CALIBRATION FOR CAR FOLLOWING: EXAMPLES

Before describing in greater detail the calibration procedure related to network models covered in Chapter 6, we will discuss some of the work on car following that appears in the literature. Car following is a key component of microscopic traffic flow models, and most of the current disaggregate calibration and validation studies were done in this area. Other calibration studies were realized in traffic flow modeling. For example, a historic paper on macroscopic models was written by Cremer and Papageorgiou (1981).

Table 4.3 lists recent studies dealing with calibration efforts. Various estimation settings have been used to solve the calibration problem, and they will be reviewed below. The studies cited in the table are classified

Table 4.3 Overview of measures of performance

Estimation method	Measure of performance	Error measure	GEH statistics	Goodness of fit function/estimator Theil's inequality coefficient	Likelihood function
Least squares (LS)	Time headway	Brockfeld et al. (2004) Punzo and Simonelli (2005)		Punzo and Simonelli (2005)	
	Intervehicle spacing	Ranjitkar et al. (2004) Punzo and Simonelli (2005) Kesting and Treiber (2008) Punzo et al. (2012)	Punzo et al. (2012)	Punzo and Simonelli (2005) Ossen and Hoogendoorn (2008a) Punzo et al. (2012)	
	Speed	Ranjitkar et al. (2004) Punzo and Simonelli (2005) Punzo et al. (2012)	Punzo et al. (2012)	Punzo and Simonelli (2005) Ossen and Hoogendoorn (2008a) Ciuffo et al. (2012) Punzo et al. (2012)	
	Speed and intervehicle spacing	Punzo et al. (2012)		Ossen et al. (2006) Ossen and Hoogendoorn (2009) Kim and Mahmassani (2011) Punzo et al. (2012)	
	Acceleration	Ossen and Hoogendoorn (2005)			

(continued)

Table 4.3 Overview of measures of performance (Continued)

Estimation method	Measure of performance	Goodness of fit function/estimator			
		Error measure	GEH statistics	Theil's inequality coefficient	Likelihood function
Maximum likelihood estimation (MLE)	Speed				Hoogendoorn et al. (2006) Hoogendoorn and Hoogendoorn (2010a) Hoogendoorn and Hoogendoorn (2010b)
	Acceleration				Ahmed (1999) Toledo et al. (2009)
Bayesian	Speed				van Hinsbergen et al. (2010)

according to the two previously mentioned and important steps of calibration: the MoP and the GoF. The remainder of this section covers the various efforts reported in the most recent literature.

4.3.1 Comparison of simulation results with observations

Three main estimation techniques have been used in the calibration of car-following models: (1) the least squares (LS) method, (2) the maximum likelihood estimation (MLE) method, and (3) the Bayesian method.

The LS method is the most widely applied technique in car-following calibration studies. A review of this approach can be found in Punzo and Simonelli (2005) and in Ossen and Hoogendoorn (2005, 2008b). Basically, the problem formulation is presented in the following equation:

$$P^* = \arg\min_{P \in D} f(Y^{obs}, Y(P)^{sim}) \tag{4.5}$$

where P is the vector of the model parameters p_i, with $i = 1,...,m$; D is the domain of feasibility of the model parameters, eventually constrained by the upper and lower bounds and by the linear and nonlinear constraints; $f(Y^{obs}, Y(P)^{sim})$ is a scalar valued nonlinear function that measures the distance between observed and simulated following behavior; and Y^{obs} and $Y(P)^{sim}$ are, respectively, the observed and simulated outputs. The domain of feasibility of the model parameters is defined by the parameter bounds and potentially by other linear and nonlinear constraints:

$$LB_i \le p_i \le UB_i \quad i = 1,...,m \tag{4.6}$$

$$g_j(P)?b_j \tag{4.7}$$

where LB_i and UB_i are, respectively, the lower and upper bounds of the parameter p_i; $g_j(P)?b_j$ is a scalar valued linear or nonlinear function of the model parameter P that calculates the left hand side of the j-th constraint; b_j is a constant value equal to the right side of the j-th constraint; and $?$ is one the $\le$, $\ge$, or $=$ relational operators.

According to this framework, the most applied estimators in the literature were the error measures. Absolute errors, square errors, percentage errors, and mixed errors were the most used. On the other hand, Theil's inequality coefficients and the GEH statistics were recently applied in calibration studies (for a comprehensive review, see Table 4.3).

The maximum likelihood estimation has also been applied in car-following model calibration. Ahmed (1999) presented the formulation of the unconditional distribution of the accelerations that constituted the like-lihood function formulation for the follower driver. Hoogendoorn et al. (2006, 2010a, 2010b) reformulated the likelihood to estimate the param-eters of a generalized form of car following model. Toledo et al. (2009) applied the method for the estimation of the parameters of an integrated lane-changing–car-following model. In a discretized form, car-following models can be expressed as follows:

$$v_i^{\text{sim}}(t_{k+1}) = \mathbb{f}\,[T, y_i(t_k), y_i(t_k - \tau)\,|\,\boldsymbol{\theta}] \tag{4.8}$$

v_i^{sim} is the simulated speed of driver i, $\boldsymbol{\theta}$ denotes the set of parameters describing the car following behavior, while T denotes the time step used for discretization. The vector $y_i(t_k)$ denotes the state that is relevant for driver i at time instant t_k. The following relation between speed data and predicted speed is assumed:

$$v_i^{\text{obs}}(t_{k+1}) = \mathbb{f}\,[T, y_i(t_k), y_i(t_k - \tau)\,|\,\boldsymbol{\theta}] + \in (t_k) = v_i^{\text{sim}}(t_{k+1}) + \in (t_k) \tag{4.9}$$

The error term $\in (t_k)$ is introduced to reflect errors in the modeling, similar to the error term used in multivariate linear regression. The error terms are generally serially correlated, as described later in this section. For now, assume that the error term is a zero mean normally distributed variable with standard deviation σ. According to the model, the difference between prediction and observation follows a normal distribution with mean 0 and standard deviation σ. The likelihood of a single prediction thus can be determined as follows:

$$\frac{1}{\sqrt{2\pi\sigma^2}}\exp\left(-\frac{\in (t_k)^2}{2\sigma^2}\right) \tag{4.10}$$

Because it has been assumed that the errors are uncorrelated, the probabil-ity of a set of observations $k = 1,\ldots, n$ can be determined, with the likeli-hood of the sample as a result:

$$L(\Theta,\sigma) = \prod_{k=1}^{n}\frac{1}{\sqrt{2\pi\sigma^2}}\exp\left(-\frac{\in (t_k)^2}{2\sigma^2}\right) \tag{4.11}$$

Applying a log transformation:

$$\tilde{L}(\Theta,\sigma) = -\frac{n}{2}\ln(2\pi\sigma^2) - \frac{1}{2\sigma^2}\sum_{k=1}^{n}\in (t_k)^2 \tag{4.12}$$

Maximum likelihood estimation involves finding the parameters that maximize the log likelihood. A necessary condition for the optimum allows determination of the standard deviation:

$$\frac{\partial \tilde{L}}{\partial \sigma^2} = 0 \Rightarrow \hat{\sigma} = \frac{1}{n} \sum_{k=1}^{n} \in (t_k)^2 \tag{4.13}$$

That is, the maximum likelihood estimate for the variance of the error is given by the mean standard error of the predictions and the observations. For the remaining parameters, the maximum likelihood estimates can be determined by numerical optimization:

$$\theta = \arg \max \tilde{L}(\Theta, \sigma) \tag{4.14}$$

with:

$$\tilde{L}(\Theta, \hat{\sigma}) = -\frac{n}{2} \left[\ln \left(\frac{2\pi}{n} \sum_{k=1}^{n} \in (t_k)^2 \right) + 1 \right] \tag{4.15}$$

This expression shows that maximization of the log likelihood is equivalent to minimization of the prediction (mean squared) error. Using the log likelihood is preferable because of its statistical properties and relevance for statistical analysis and cross-comparison. However, subsequent error terms in trajectory data are not independent and show serial correlation or autocorrelation. Indeed, the covariance between the subsequent errors exceeds zero:

$$\rho = \text{cov}(\in_k, \in_{k+1}) > 0 \tag{4.16}$$

A review of the approach to deal with serial correlation can be found in Hoogendoorn and Hoogendoorn (2010a, 2010b). They proposed the following three-step approach to determine the extent to which serial correlation plays a role: (1) obtain the parameter estimates by optimization of the log-likelihood; (2) determine the errors; and (3) determine ρ.

The Durbin-Watson test can then be applied to determine whether the estimate of the autocorrelation coefficient significantly differs from zero. If the correlation coefficient $\rho \neq 0$, it is necessary to transform the model to determine correct values for the parameter variance estimates. To approximate the covariance matrix of the estimated parameters, the so-called Cramér-Rao lower bound can be used:

$$\text{var}(\hat{\theta}) \geq -E(\nabla^2 \tilde{L}) \tag{4.17}$$

For the maximum likelihood estimators, the asymptotic variance of the parameters is given by the right side of the previous equation.

The Bayesian approach is a generalization of the likelihood ratio test (LRT) introduced in Hoogendoorn et al. (2006). To test whether one model performs better than another, the likelihood ratio test is performed. To this end, the zero acceleration model is used as a reference:

$$v_i^{\text{obs}}(t_{k+1}) = v_i^{\text{sim}}(t_k) + \in (t_k) \tag{4.18}$$

For this model, one can determine the (null) log likelihood:

$$\tilde{L}_0 = -\frac{n}{2}\left[\ln\left(\frac{2\pi}{n}\sum_{k=1}^{n}\in(t_k)^2\right)+1\right] \tag{4.19}$$

The LRT involves testing the statistic:

$$2[\tilde{L}(\Theta,\hat{\sigma}) - \tilde{L}_0] \tag{4.20}$$

which follows a χ^2 distribution with degrees of freedom equal to the number of model parameters to calibrate. The LRT is passed with $(1-\alpha)$ confidence if:

$$2[\tilde{L}(\Theta,\hat{\sigma}) - \tilde{L}_0] > \chi^2(1-\alpha,d) \tag{4.21}$$

The likelihood ratio test can also be used to cross-compare the performances of two car-following models. In this case, d denotes the difference in the number of parameters of the complex model and the simple model. The test accounts for the number of parameters (via degrees of freedom d) and thereby makes it possible to compare simple and complex models correctly.

In the Bayesian method, prior probabilities are transformed into posterior probabilities for each parameter in the car-following model for which Bayes' rule is used. The exact formulation of this method for calibration and model selection is presented in van Hinsbergen et al. (2010). This approach has several advantages over existing mechanisms.

The most important feature is that it leads to a probabilistic approach to compare different models on the basis of posterior distributions of their parameters. This allows selection of a model that probably best describes a certain driver's behavior, taking into account both the calibration error and the model complexity. The main contribution of the Bayesian approach compared to the LRT is that any model that is differentiable to its parameters can be used.

Just as with the LRT approach, prior information can be included when calibrating the parameters of car-following models to rule out unrealistic estimation results arising from inadequate information about certain parameters.

The approach can be used to combine the predictions of several models in a so-called committee or ensemble in which different models predict the behavior of one single driver, which may help decrease errors due to intra-driver differences. Error bars can be constructed from the predictions of car-following models.

4.3.2 Optimization methods

Given the earlier reported problem specifications, three main optimization algorithms have been used in the field of car-following model calibration to find the model parameter estimates, specifically: (1) downhill simplex, (2) genetic algorithm (GA), and (3) OptQuest/Multistart.

The downhill simplex, or Nelder–Mead, method was proposed by John Nelder and Roger Mead (1965). It is a gradient-free optimization method, widely used in many car following model calibration studies since 2004 (Brockfeld et al., 2004 and 2005; Ossen et al., 2006, 2008, and 2009; Kim et al., 2011).

It is a common unconstrained nonlinear optimization technique since it is a well-defined numerical method for twice differentiable problems. However, the Nelder–Mead method is only a heuristic, since it can converge to nonstationary points (Powell, 1973; Lagarias et al., 1998; McKinnon, 1996) on problems that can be solved by alternative methods.

Since the algorithm does not allow the setting of bounded or constrained parameters, most car-following calibration studies add a penalty value to the objective function value if parameter values are not defined in the domain of feasibility constrained by the upper and lower bounds and by nonlinear constraints (see problem specification in discussion of least squares method).

GAs are widely used for the calibration of microscopic traffic simulation models. The reason is clear: no information on the objective function is required for their application and they are thus suitable for black-box optimization. For the calibration of car following models, they have been applied by, for example, Ranjitkar et al. (2004), Kesting and Treiber (2005), and Punzo et al. (2012).

Even though GAs are suitable for solving constrained nonlinear optimization problems, only the parameters bounds were set in designing the problem since it was recognized that nonlinear constraints greatly slowed the optimization. Indeed, the GA uses the augmented Lagrangian genetic algorithm (ALGA) to solve nonlinear constrainted (and bounded) problems.

With this approach, bounds and linear constraints are handled separately from nonlinear constraints. Thus, a subproblem is formulated by combining the fitness function and nonlinear constrainted function using the Lagrangian and penalty parameters. A sequence of such optimization problems is approximately minimized using the GA such that the linear constraints and bounds are satisfied. As a result, the algorithm minimizes a sequence of the subproblem, which is an approximation of the original problem, resulting in an increase of the number of function evaluations needed to solve it (Conn et al., 1991, 1997). Thus, to limit the computing time, a penalty fraction is often applied to simulate the nonlinear constraints of the domain of feasibility.

Since the GA is not a global optimizer, it may present difficulties in finding a stationary global solution. However, the GA can sometimes overcome this deficiency with the right settings. Indeed, with a large population size, the GA searches the solution space more thoroughly, thereby reducing the chance that it will return a local minimum that is not global (Powell, 1973). Concurrently, a large population size also causes the algorithm to run more slowly.

The OptQuest/Multistart heuristic (Ugray et al., 2005) is an optimization algorithm for solving both constrained and unconstrained global optimization problems. It has been recently used for the calibration of car following models in Punzo and Simonelli (2005) and in Punzo et al. (2012).

For a more detailed description please refer to annex b.

In spite of the large number of studies attempting to expand the properties of models and phenomena through the results of calibrations, few attempted also to analyze and quantify the uncertainties involved in the calibration process and their impacts on the accuracy and reliability of results. For example, Brockfeld et al. (2004) recognized that many optimization algorithms are stuck in local minima and suggested starting the algorithms at least five times from different starting points, as also indicated in Ossen et al. (2006).

Punzo and Simonelli (2005) pointed at the effect on calibration results of using different MoPs in the objective function, namely, speed, intervehicle spacing, and headway, providing numerical comparisons and a justification of the advantage of using spacing. Kesting and Treiber (2008) confirmed the justification of Punzo and Simonelli for preferring intervehicle spacing and compared the effect on results of using different GoF functions such as relative error, absolute error, and mixed error.

Finally, Ossen and Hoogendoorn (2008a, 2009) asserted the preeminent role of experiments with synthetic data to investigate calibration issues and showed that measurement errors can yield considerable bias in estimation results. They also raised the crucial issue that parameters minimizing the objective function do not necessarily best capture following dynamics, and, as a general conclusion, they stated that "calibration based on real trajectory data turns out to be far from trivial."

Chiabaut et al. (2010) recently approached the calibration problem starting from the observation of macroscopic patterns in congestion (i.e., propagation of waves generated by the model) and attempted to calibrate the parameters of the stochastic Newell car-following model. The objective was to find the uniform coordinates of the shock wave speed that minimizes the discrepancy of the measurements (assessed by the standard deviation). Therefore, the GoF function is embedded in the variation of the measured coordinates of the shock wave speed. This criterion appeared not to be flat like the classic error measures and Theil's inequality coefficients. Moreover, is robust to measurement errors and no multiple qussi-optimal solutions can corrupt the final result.

Analysis of complete chain MoP and GoF measurement optimization

Punzo et al. (2012) attempted to shed more light on the calibration of car following models against trajectory data. They presented the main findings of a vast exploratory study aimed at investigating and quantifying the uncertainty entailed in calibration processes. All the combinations of algorithms, MoPs, and GoF functions applied in the literature in the past 10 years were tested.

Each test was performed several times from different starting points to reveal the impacts of the initial settings on the calibration results. The verification approach was based on experiments with synthetic data. This allowed the researchers to ascertain whether and how well optimal parameters were identified by the calibration procedure (Ossen and Hoogendoorn, 2008a). Several outcomes resulted from the study as described below.

GoF functions based on the GEH statistics were affected greatly by the setting of the threshold value. When used in calibration, a wrong setting of this value led to the loss of uniqueness of the global solution, even in the case of optimization problems on synthetic data where the global minimizer is unique and well-defined.

The downhill simplex was unable to rediscover the true set of the parameter values in any of the experiments. Further, the heuristics was very sensible to the initial starting condition, providing very different sets of optimal parameters based on the starting point. Thus, results obtained by a single-shot calibration were neither optimal nor meaningful.

Both the GA and the OptQuest/Multistart were able to find the "known" global minimizer at least once over 64 replications of the same calibration experiment. Moreover, they were able to rediscover the true values of the most sensitive parameters in most replications.

The use of mixed GoF functions combining both MoPs (speed and spacing) such as the sum of Theil's inequality coefficients performed worse than GoFs calibrated separately on speed or on spacing. Further, the use of absolute measures of the distance between observed and (model) simulated outputs such as the MAE yielded very low efficiency in the optimization because they require high numbers of evaluations of the objective functions to satisfy the same stopping rules adopted with other GoF functions. Moreover, the improvements in finding the global minimizer were negligible.

As a general conclusion, the study confirmed the complexity of the problem of calibrating car-following models against real trajectory data and provided the research lines for building more robust calibration settings.

4.4 IMPORTANCE OF USING SYNTHETIC DATA FOR COMPLETE CALIBRATION PROCEDURE TESTING

We have seen in this chapter that many choices are involved in the calibration and validation processes. MoPs must be chosen to agree with the assessment process within which the simulation tool is used. If one wants to evaluate the impact of a modification of an intersection on total time spent, travel time must be one of the MoPs chosen. Other values are GoFs and optimization algorithms. All these choices and others (that will be described in Chapter 6) significantly inflate the resulting optimal parameters and validation results.

Because heuristic optimization processes such as the ones presented in Annex B are not guarantees of finding an optimum near enough to a real optimal value, it is desirable to evaluate the correctness of the calibration process. To this end, using synthetic data with a known optimal parameter set to test the procedure is a good way of proceeding.

After "realistic" inputs to a simulation model (time-varying OD matrices, realistic parameter values, etc.) have been chosen, the model is run and outputs from several replications (each using a different seed in the model random numbers generation process) are averaged to reduce the effects of stochasticity, thus producing synthetic "true" measurements. The parameters used in the simulation are therefore the "true" parameters. A noise is added to the synthetic "true" measurements in order to mimic the measurements errors.

Black-box calibration then tries to find parameter values that minimize the distance (measured by a GoF) between true and simulated MoPs. The calibration verifies whether the calibrated parameters equal the "true" ones at the end of the process. If the parameters are not equal, the calibration is repeated by changing one of the variables and possibly also the sensitive parameters considered. After the procedure is verified, it can be applied using the observed measurements. Many examples of the procedures may be found in the recent literature.

To the best of our knowledge, the first discussion of this method for microscopic simulation was in Ossen (2008a). Another example can be found in Punzo et al. (2012).

They presented the main findings of a vast exploratory study aimed at investigating and quantifying the uncertainty entailed in calibration processes. All the combinations of algorithms, MoPs, and GoF functions applied in the literature in the past 10 years were tested.

Each test was performed several times from different starting points to reveal the impacts of the initial settings on the calibration results. The verification approach was based on experiments with synthetic data. This allowed the researchers to ascertain whether and how well optimal parameters were identified by the calibration procedure like in the previous reference (Ossen and Hoogendoorn, 2008a). Several outcomes resulted from the study as described below.

GoF functions based on the GEH statistics were affected greatly by the setting of the threshold value. When used in calibration, a wrong setting of this value led to the loss of uniqueness of the global solution, even in the case of optimization problems on synthetic data where the global minimizer is unique and well-defined. The downhill simplex was unable to rediscover the true set of the parameter values in any of the experiments. Further, the heuristics was very sensible to the initial starting condition, providing very different sets of optimal parameters based on the starting point. Thus, results obtained by a singleshot calibration were neither optimal nor meaningful.

Both the GA and the OptQuest/Multistart were able to find the "known" global minimizer at least once over 64 replications of the same calibration experiment. Moreover, they were able to rediscover the true values of the most sensitive parameters in most replications.

The use of mixed GoF functions combining both MoPs (speed and spacing) such as the sum of Theil's inequality coefficients performed worse than GoFs calibrated separately on speed or on spacing. Further, the use of absolute measures of the distance between observed and (model) simulated outputs such as the MAE yielded very low efficiency in the optimization because they require high numbers of evaluations of the objective functions to satisfy the same stopping rules adopted with other GoF functions. Moreover, the improvements in finding the global minimizer were negligible.

As a general conclusion, the study confirmed the complexity of the problem of calibrating car-following models against real trajectory data and provided the research lines for building more robust calibration settings.

4.5 CONCLUSIONS

Traffic simulation models are typically used for the assessment and planning of road infrastructures, the evaluation of advanced traffic management and information strategies, and the testing of technologies and systems intended to increase the safety, capacity, and environmental efficiency of vehicles and roads. For each study, traffic phenomena and behaviors relevant to the study are selected and should be described accurately by a simulation model. Preferably, this detailed analysis is performed and reported by the developers of the simulation tool.

In addition, the calibration and validation should be completed for the application scenario.

Globally, calibration and validation of a simulation tool are the processes of comparing the simulation results for a chosen variable(s) with a set of observations of this variable(s). Figure 4.1 shows a generic framework for the calibration and validation processes. In calibration, the differences between the simulated and observed variable(s) are minimized by finding an optimum set of model parameters. To define the parameters to which the calibration procedure is to be applied, the reader is referred to Chapter 5 covering sensitivity analysis.

The practical specification of calibration and validation relies on the error function (measure of GoF), traffic measurement (MoP), traffic simulation model used, transportation system (traffic scenario) to be simulated, demand pattern used and the accuracy of its estimation, parameters to calibrate, quality (error presence or absence) of the observed measurement, and possible constraints. This chapter has shown a variety of measures of performance (MoP) and goodness (GoF) of fit and discussed the advantages and disadvantages of the variable choices.

Chapter 5

Sensitivity analysis

Vincenzo Punzo, Biagio Ciuffo, and Marcello Montanino

CONTENTS

5.1 INTRODUCTION

"What makes modeling and scientific inquiry in general so painful is uncertainty. Uncertainty is not an accident of the scientific method, but its substance" (Saltelli et al., 2008).[1] To understand how uncertainty enters traffic modeling it is useful to consider the sources and nature of uncertainties in traffic systems models.

---

[1] "That is what we meant by science. That both question and answer are tied up with uncertainty, and that they are painful. But that there is no way around them. And that you hide nothing; instead, everything is brought out into the open" (Høeg, 1995).

The trajectories of vehicles that fully depict the evolution of traffic over a road network are the outcomes of a number of human choices and actions. For the sake of simplicity, it is generally acknowledged that a number of decisions such as the time to depart or the route to follow are made at a driver's "strategic" decision level. A "tactical" level involves decisions and actions aimed to directly control a vehicle in a traffic stream, subject to a number of environmental constraints (road rules, traffic lights, surrounding traffic) and to drivers strategic plans and motivations.

Traffic simulation aims to reproduce traffic over road networks by more or less explicitly modeling these strategic and tactical decision layers. In a microscopic approach to simulation, for instance, most of the decisions at both levels are explicitly modeled for each individual vehicle. This composite modeling process involves a number of uncertainty sources of different natures, often mixed in complex ways. Part of this uncertainty can be directly imputed to the adequacy or inadequacy of a model in relation to reality while another part depends on the uncertain model inputs.

Uncertainty due to the inadequacy of models arises from a number of sources like the basic modeling assumptions, the structural equations, the level of discretization, and the numerical resolution method. Uncertainties may be reduced by improving one or more of these aspects of a model. Because the cost of reducing uncertainties often increases computing time, the choice of the most appropriate modeling framework depends on the specific application (e.g., on-line versus off-line simulation) and stems from a trade-off between model adequacy and computing time.

With regard to input uncertainty, we must distinguish inputs that are observable from those that are not. This distinction is crucial because it affects the possibility and/or the cost of reducing the uncertainties for which inputs are responsible:

- Observable inputs have measurable equivalents in reality. Thus, they can be directly estimated, i.e., measured, and used to feed the models. In a traffic microscopic model, examples are the network characteristics such as traffic light timing, traffic composition, and distribution of vehicle sizes.
- Unobservable inputs are those that either are hardly measurable[2], like the OD demand, or that have no actual equivalent in the reality, like the model constants. Such inputs can be therefore only indirectly estimated by means of inverse analysis, calibration, and other techniques (see section 5.2).

Many model parameters are in-between these two categories. Though they have a physical interpretation, often they are indirectly estimated as if

---

[2] In this context the immeasurability is intended practical rather than theoretical. Some quantities may be not measurable because of operational or economic constraints.

they were unobservable. In facts, as traffic models are necessarily only coarse representations of the real system, considering model parameters as uncertain quantities to be indirectly estimated is commonly taken to cover both the uncertainty regarding the un-modelled details of the phenomena (the so-called epistemic uncertainty) and the aleatory component not predicted by the average models e.g. the variability in time of driver's behaviour.[3]

Table 5.1 provides examples of uncertainty sources in traffic modeling, classified by their natures. The distinction is based on the practical notion of reducibility rather than on theoretical distinctions like epistemic versus aleatory. It is worth noting that the table gives general indications, but, depending on the model and application context, a source may be classified in different ways. A typical example is whether to include model uncertainty with the uncertainties in parametric inputs.

Since the uncertainties in both the model and the inputs are propagated into the outputs, such uncertainties must be assessed and, when possible, reduced. In fact, a model encompassing a disproportionate amount of uncertainty and thus returning unreliable results has no practical utility for a transport analyst. Thus, a conceptual framework generally adopted for uncertainty assessment and management is depicted in Figure 5.1 (see de Rocquigny et al., 2008).

The first step (A in the figure) consists of problem specification, which involves the definition of the input and output variables, model specification, and the identification of the quantity of interest for measuring the uncertainties in the outputs. Input variables may be uncertain (x) or fixed (d), based on the choice of the analyst. Depending on the problem setting, the uncertain x model inputs may include all the sources like the parametric and model uncertainties. Other variables may be fixed (and denoted with d), for example, in risk scenarios d for comparative studies or more generally when the uncertainty of these variables are deemed negligible with respect to the output variables of interest.

The second step (B) is the quantification of the uncertainty sources (uncertainty modeling). In a probabilistic setting, this phase implies defining the joint probability density function (pdf) of the uncertain inputs or their marginal pdf values with simplified correlation structures or even independence assumptions. The step involves gathering information via direct observations, expert judgments, physical arguments, or indirect estimation (as for the unobservable inputs in Table 5.1). Step B is often the most expensive phase of an analysis.

---

[3] Epistemic, or reducible uncertainty, refers to types of uncertainty which can be directly reduced by an increase in available data. Aleatory, or irreducible uncertainty, refers to events which remain unpredictable whatever the amount of data available. The difference is clear when looking at the failure rate of an industrial component (epistemic) against its instant of failure (aleatory).

Table 5.1 Example of uncertainty sources in traffic modelling and their nature

Uncertainty source		Uncertainty nature		
		Mostly reducible	Mostly irreducible	Mixed natures
Model		Time and space discretization	Model time-invariance	Basic modeling assumptions
Observable inputs		Road characteristics and functions; traffic control states; whole traffic composition; point to point demand (e.g. on freeway network), etc...		Vehicle sizes; free flow speeds, etc. (variability in population/space)
Unobservable inputs	Hardly measurable	Stationary OD matrices	Individual departing time	Time-varying OD matrices
	Unphysical parameters	Aggregate model constants		Disaggregate model constants
	Uncertain parameters	Fundamental diagram parameters (jam density, etc.), cost coefficients, etc.	Variability in time of parameters (like the driver reaction time)	Reaction times; maximum acceleration/ decelerations; desired speeds, etc. (variability in population)

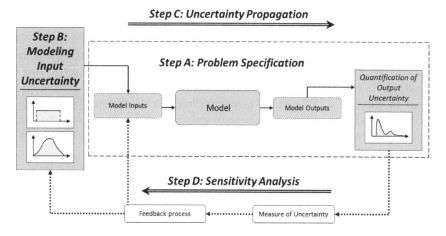

Figure 5.1 Generic conceptual framework for uncertainty management.

The propagation of the uncertainty (Step C) is necessary to map the uncertainty in the inputs into the uncertainty measures in the outputs. A Monte Carlo simulation framework is often adopted. In such a probabilistic framework, propagation entails the estimation of the pdf of the output variables of interest y, given the pdf values of the uncertain inputs x, the values of the fixed inputs d, and the model $f(x,d)$.

The sensitivity analysis (SA) or importance ranking (Step D) represents the feedback process in the complex of the uncertainty management. Its purpose is understanding "how uncertainties in the model outputs can be apportioned to different sources of uncertainties in the model inputs" (Saltelli et al., 2004). In other words, "the objective of the sensitivity analysis is to instruct the modeler with regard to the relative importance of the uncertain inputs in determining the variable of interest." It generally involves some statistical treatment of the input-output relations drawn within the uncertainty propagation step, which will be introduced in Section 5.3. Some considerations about the role of SA in traffic modeling are reported in the next section.

5.2 SA APPROACHES IN TRAFFIC MODELING

As noted in the introduction, whatever the degree of complexity, a mathematical model is still a simplified representation of a real traffic phenomenon. Thus, it is common practice to consider model uncertainty alongside the parametric inputs. Calibrating the uncertain model parameters against real-world outputs allows covering in one operation the uncertainty in the system or phenomenon, the errors in the data, and the inaccuracies of the model.

To give an example, estimating the probability distribution function (pdf) of the parameters of a car-following model is needed to account for the variability of driving behaviors within the population (inter-driver variability) and also to compensate the model errors and aleatory components such as the variability in time of driving behaviors (intra-driver variability). This is also the theoretical reason why such model parameters must be estimated indirectly—using the model rather than directly estimating them from real world measurements, even when they have physical equivalents in reality. The pdfs resulting from the two approaches are different indeed.

The problems with reducing parametric input uncertainty via inverse analysis methods such as indirect estimation or calibration mainly concern five factors:

- The scarceness, incompleteness, or inconsistency of data in relation to model complexity
- The data measurement errors
- The computational complexity of the estimation process
- Set-up suitable to the nature of the specific problem
- The asymmetry in the importance of uncertain inputs

The scarceness, incompleteness, or inconsistency of data with respect to the complexity of a model can lead to ill-posed inverse problems such as in the static OD matrix estimation problem (Marzano et al., 2009) or to biased or not robust estimates of the parameter pdfs. The latter effect also arises in the presence of measurement errors (Ossen and Hoogendoorn, 2008a).

Large numbers of parameters can make an analysis computationally unfeasible. For example, in least squares black-box calibration of model parameters, the computational complexity is exponential because of the number of parameters, making the search for a global optimum generally unfeasible even for a relatively small number of parameters (Ciuffo et al., 2008).

The problem set-up, that includes the choices of algorithm, MoP, and GoF function in the objective, sensibly affects the quality of the solution. In particular, it has influence on the chance of finding a global optimum or at least a stable solution. Punzo et al. (2012) presented an exploratory study on this subject (see Section 4.6.3). In addition most models present a pronounced asymmetry in the influence of the parametric inputs on their outputs, with a small subset of input parameters accounting for most of the output uncertainty and the others playing little or no role. The calibration of parameters with scarce influence on the outputs along with flat objective functions, for instance, is a challenge for any optimization algorithm.

The key role played by SA may serve a number of useful purposes, depending on the specific setting adopted. The importance ranking of the inputs with regard to their influence on the output uncertainty (factor prioritization setting) is the most common function of SA. The analysis can be used to identify which input parameters really need to be calibrated (factor fixing setting) and which are the observations are really sensitive to the inputs and thus useful for the estimation. Reducing the number of parameters to calibrate may make feasible an otherwise unfeasible problem while the definition of the most appropriate observations is crucial for guiding the allocation of resources for the collection of new data. Both tasks are obviously crucial to reducing costs and producing a successful analysis.

In addition to the importance of ranking of uncertainty sources, SA may be useful for identifying the elements of a modeling process (inputs, assumptions, etc.) and the regions of the inputs that are most responsible for yielding acceptable model realizations or, conversely, exceeding specific thresholds (i/o mapping settings).

For all these reasons, SA may be viewed as a formalized and efficient space-exploring mathematical tool for investigating upstream models. It is not surprising that SA may uncover technical errors in a model or aid simplification of a model. Often this analysis helps modify the uncertainty model or causes system or design modifications to reduce uncertainty.

Despite the importance of SA in scientific modeling, few examples exist in the traffic modeling literature. These are reported in the next section along with a brief explanation of the corresponding techniques applied. SA techniques are generally classified (Frey and Patil, 2002) as:

- Graphical methods for screening sensitivity analysis
- Mathematical methods for local sensitivity analysis
- Statistical methods for global sensitivity analysis, based on design of experiment (DoE)

In the first group, graphical methods, represent sensitivity in the form of graphs, charts, or surfaces. Generally, graphical methods are used to indicate visually how an output is affected by variations in inputs. A graphical method can be used as a screening step before further analysis of a model. These methods are frequently used to complement the results of mathematical and statistical methods for better representation. Input-Output scatter plots are the most frequently used (see Section 5.3.1). Other graphical representations are spider plots, tornado plots, and pie plots (Ji and Prevedouros, 2006).

In mathematical methods, the effects of model inputs (parameters) on outputs are evaluated individually; that is, the variation in the model outputs created by the variation of one input at a time is studied while all other inputs are kept at fixed values. This approach has two major drawbacks. As inputs never vary simultaneously, the method completely hides the interaction effects of the parameters and thus provides unbiased results only for purely additive models (never the case of traffic models). The second drawback is that the method is local by nature: it investigates the neighborhood of a specific point and is not informative about the rest of the input space. In nonlinear models, where the results cannot be extrapolated simply in other points of the space by linear interpolation, the analysis results are necessarily of partial validity (i.e., local). The one-at-time sensitivity analysis (Section 5.3.2), the elementary effect test (Section 5.3.3), and the sigma normalized derivatives (Section 5.3.4) belong to this group.

In the third group, global sensitivity analysis methods are used to apportion the uncertainty in the outputs to different sources of uncertainty in the input variables. These analyses are based on statistical methods that involve the running of several model simulations in which the model input parameters are drawn from estimated probability distribution functions according to a specific DoE.

The DoE or setting used to explore the range of parameter values can be done using a variety of techniques such as full factorial design, fractional factorial design, Monte Carlo sampling (MCS), Latin hypercube sampling (LHS), quasi–Monte Carlo sampling, and several others. This kind of experimental setting allows a global exploration of a whole range of parameter pdfs and the identification of interaction effects among multiple model inputs. For these reasons, these methods are known as global sensitivity analyses.

Regression and correlation analysis and variance-based methods belong to this group. Regression and correlation analysis involves random generation of the model parameters according to a specific DoE and repeated execution of the model. Regression or correlation analysis is then performed between dependent and independent variables, which derives the standardized regression coefficient analysis (Section 5.3.5) and the partial correlation coefficient analysis (Section 5.3.6).

The absolute values of regression coefficients and correlation coefficients indicate qualitatively how much the corresponding uncertain inputs contributed to the uncertainty of the output. These methods are straightforward to understand and easy to apply since they are often performed together with MCS. However, they generally adopt linear regression models and therefore are not able to capture the higher-order interaction effects between model inputs and outputs.

Conversely, the variance-based methods are used to decompose the variances of outputs into different sources attributable to each model factor (input parameter) to account for the first-order (linear) effects and the interaction (higher-order) effects among inputs. Because of constraints on the analytical feasibility of the functional forms of traffic simulation models, these methods are implemented through model simulation according to a DoE. Among the other techniques, the most commonly used variance-based methods are the ANOVA (Section 5.3.9), the Fourier amplitude sensitivity test (Section 5.3.10), and the methods based on the Sobol decomposition of variance (Section 5.3.11).

Beyond the approaches described above, the *sensitivity analysis* term has been used in several other works to identify analyses that do not belong to the general framework presented in Figure 5.1 and for this reason are not detailed here. Often such analyses are local in nature, unlike the global approach depicted earlier and for this reason will not be reported in the following review of existing works on SA.

5.3 BACKGROUND ON TECHNIQUES AND APPLICATIONS IN TRAFFIC MODELING

This section reviews some of the most common SA techniques and their applications in the field of traffic modeling. In particular, the variance-based methods based on the Sobol decomposition of variance will be covered in more detail because they are recognized as meaningful tools for performing global sensitivity analysis. The main source for information is still Saltelli et al. (2008) whose work provided most of the relevant material and to which the reader is directed for an in-depth analysis.

Several techniques have been utilized for analyzing the sensitivity of a model. Among others, we cite (1) input and output scatter plots, (2) one-at-time (OAT) sensitivity analysis, (3) the elementary effect test, (4) the sigma-normalized derivatives, (5) the partial correlation coefficient analysis, (6) the standardized regression coefficient analysis, (7) Monte Carlo filtering, (8) metamodeling, (9) factorial analysis of variance (ANOVA), (10) the Fourier amplitude sensitivity test, and (11) the variance-based method based on the Sobol decomposition of variance. Some elements of each technique are presented.

5.3.1 Input and output scatter plots

Let the model considered be in the form

$$Y = f(Z_1, Z_2, ..., Z_r) \tag{5.1}$$

where Z_i $(i = 1, ..., r)$ is the model input and Y its output. We now perform a Monte Carlo experiment with our model. This means that, based on the statistical distribution of the model inputs, we sample N possible combinations of them to generate the following matrix:

$$
M = \begin{bmatrix}
z_1^{(1)} & z_2^{(1)} & \cdots & z_r^{(1)} \\
z_1^{(2)} & z_2^{(2)} & \cdots & z_r^{(2)} \\
\cdots & \cdots & \cdots & \cdots \\
z_1^{(N)} & z_2^{(N)} & \cdots & z_r^{(N)}
\end{bmatrix}
\tag{5.2}
$$

Computing Y for each row of the matrix in Equation (5.2), we obtain the vector of model outputs Y.

$$
Y = \begin{bmatrix}
y^{(1)} \\
y^{(2)} \\
\cdots \\
y^{(N)}
\end{bmatrix}
\tag{5.3}
$$

If we now plot the elements of Y against the correspondent elements of each column of M, we obtain r scatter plots. From the visual analysis of the various plots, it is possible to identify the parameters that exert an influence on the model outputs and parameters that do not. For the parameters that may influence the model outputs, the cloud of points of the scatter plot will have a more or less defined shape. For the others, the cloud will approximately resemble a circle. In this way, the scatter plot represents the simplest way to perform sensitivity analysis. The problem is that the method becomes impractical if the number of variables increases. In addition, it does not allow for the investigation of the sensitivities of groups of variables.

5.3.2 One-at-time (OAT) sensitivity analysis

In the OAT sensitivity analysis, we study the variations in model outputs due to the variation of one input parameter at a time, while the remaining parameters are fixed at certain values. The difference in model output due to the change in the input variables is known as the sensitivity or swing weight of the model to that particular input variable (Morgan and Henrion, 1990).

The approach has been applied with the VISSIM model by Lownes and Mechemel (2006) and Mathew and Radhakrishnan (2010), respectively, to prioritize model parameters in terms of their effects on model outputs and select the parameters to be calibrated. The effect of travel demand variability on travel times was investigated by Bloomberg and Dale (2000) using the VISSIM and CORSIM micro-simulation software. These screening sensitivity analyses were also performed by Lawe et al. (2009) to assess the sensitivity of the TRANSIMS model results (traffic volumes and average speeds) to changes in the random seed number and the pretimed signals to actuated controllers.

Kesting and Treiber (2008) followed the same approach to gain additional insight about the meanings of the values of parameters resulting from the calibration of two car-following models. In Patel et al. (2003), the authors focused on the uncertainties in input data for the CAL3QHC roadway emission model and applied OAT analysis to identify the most sensitive parameters of the model-simulated carbon monoxide concentrations.

However, as remarked in the previous section, this approach is "illicit and unjustified unless the model under analysis is proved to be linear" (Saltelli et al., 2006). Indeed, in this study, they show that available good practices such as variance-based measures and others, are able to overcome OAT shortcomings and are easy to implement. These methods also allow factors importance to be defined rigorously, thus making the ranking univocal.

5.3.3 Elementary effect test

The test basically consists of an average of derivatives over the space of factors. If the r input variables vary across p levels, the elementary effect of the i-th input variable at level j is:

$$EE_{i,j} = \frac{Y(Z_1,\ldots,Z_i + \Delta_j,\ldots,Z_r) - Y(Z_1,\ldots,Z_i,\ldots,Z_r)}{\Delta_j} \qquad (5.4)$$

Δ_j is the width of the level j. The sensitivity index for the i-th variable is then evaluated:

$$\mu_i = \frac{1}{p}\sum_{j=1}^{p} |EE_{i,j}| \qquad (5.5)$$

which allows the variables to be ranked. In this way, the elementary effect test can be considered a screening method, preferably used before the application of a more sophisticated method to reduce the number of input variables to consider.

To the best of the authors' knowledge, no studies have ever applied this technique in transportation modeling and simulation. For a teaching application in a different research field, the interested reader may refer to Campolongo et al. (2007).

5.3.4 Sigma normalized derivatives

Function's derivatives seem the most natural way to perform sensitivity analysis, especially for analytical models. In truth, derivatives are not always suitable (Saltelli et al., 2008) for this purpose and sigma normalized derivatives are used instead. Considering the previous example, the formulation for sigma normalized derivatives is:

$$S_{Z_i}^{\sigma} = \frac{\sigma_{Z_i}}{\sigma_Y} \frac{\partial Y}{\partial Z_i} \tag{5.6}$$

in which $S_{Z_i}^{\sigma}$ represents the sensitivity index for the variable Z_i and σ the standard deviation. It is worth noting that a sensitivity index as in Equation (5.6) is recommended for sensitivity analysis by the Intergovernmental Panel for Climate Change (IPPC).

The main shortcoming of this approach is in application with black-box models (i.e., simulation-based). In this case, the derivative computation can be very time consuming and thus expensive. For this reason, derivatives are usually evaluated only in the middle of the distribution of the single variables and some hypotheses on the function are made to extrapolate results to the entire function. When the hypotheses prove false, the results achieved may be misleading.

In recent years, this technique has been applied more than once without the sigma normalization. In the field of traffic demand assignment, it was used to perform a sensitivity analysis of equilibrium network flows by Tobin and Friesz (1988) and further extended by Yang (1997). In the same context, Leurent (1998) performed an analysis of the dual criteria traffic assignment model, while Yang (1998) investigated the sensitivity analysis of the queuing equilibrium network flow to derive explicit expressions of the derivatives of equilibrium link flows and equilibrium queuing times for traffic control parameters. Later, the same approach was adopted by Chen et al. (2002) to estimate the sensibility of the network equilibrium flows to the change of arc capacities.

In the context of traffic simulation, the sigma normalized derivatives have been effectively used by Ji and Prevedouros (2005a) to perform a sensitivity analysis of the delay model proposed in the HCM (2000). In the same study, the authors compared most of the other techniques that will be presented in the following sections.

5.3.5 Standardized regression coefficient (SRC) analysis

Another possibility for non-black-box models is to create a regression model on the basis of the evaluations of a function. If we consider again elements of Equations (5.2) and (5.3), a linear regression model can be written in the form:

$$y^{(j)} = b_0 + \sum_{i=1}^{r} b_{Z_i} Z_i^{(j)} \tag{5.7}$$

in which bz_i represents the coefficients of the regression model. Normalizing these coefficients with the standard deviations of input and output, we obtain the sensitivity index:

$$\hat{\beta}_{Z_i} = \hat{b}_{Z_i} \frac{\sigma_{Z_i}}{\sigma_Y} \tag{5.8}$$

For linear models, the sensitivity index in Equation (5.8) coincides with that of Equation (5.6). This holds only in this case. In general, standardized regression coefficients are more robust and reliable than sigma normalized derivatives as they result from exploration of the entire spaces of the input variables. However, their precision depends on the size N of the Monte Carlo experiment. To the best of the authors' knowledge, this technique was applied in the context of traffic modeling only by Ji and Prevedouros (2005a and 2006) on the delay model proposed by the HCM (2000).

5.3.6 Partial correlation coefficient (PCC) analysis

A simple method for assessing the relationship between independent and dependent variables is calculation of the correlation coefficient for the values of input parameters and the output. The most frequently used method for linear correlation is the Pearson product moment correlation coefficient (R) expressed as:

$$R = \frac{\sum_{i=1}^{n} x_i y_i - \frac{1}{n}(\sum_{i=1}^{n} x_i)(\sum_{i=1}^{n} y_i)}{\sqrt{\sum_{i=1}^{n} x_i^2 - \frac{1}{n}(\sum_{i=1}^{n} x_i)^2} \sqrt{\sum_{i=1}^{n} y_i^2 - \frac{1}{n}(\sum_{i=1}^{n} y_i)^2}} \tag{5.9}$$

where:
   x_i is an observation of the model input X, $i = 1,...,n$.
   y_i is an observation of the model output Y, $i = 1,...,n$.
   n is the number of observations of the model inputs and outputs.

The drawback of this simple function is that a strong correlation between input parameters may influence input–output correlations. The partial correlation coefficient is calculated to represent the linear relationship between two variables with a correction to remove the linear effects of all other variables. For example, given a dependent variable Y with two independent variables X_1 and X_2, the partial correlation coefficient between X_1 and Y eliminates the possible indirect correlations between X_2 and Y and $(X1,X_2)$ and Y. PCC can be defined as:

$$R_{X_1 Y | X_2} = \frac{R_{X_1 Y} - R_{X_1 X_2} R_{X_2 Y}}{\sqrt{(1 - R_{X_1} X_2^2)(1 - R_{X_2} Y^2)}} \tag{5.10}$$

where $R_{X_1Y|X_2}$ represents the PCC for X_1 and Y while removing the effects of X_2. This technique has been applied sporadically in the context of traffic modeling and simulation.

An exhaustive application can be found in Ji and Prevedouros (2005a) who applied the technique to perform a sensitivity analysis of the delay model proposed in the HCM (Highway Research Board, 2000). The authors also reviewed other methods introduced here and compared their performances to evaluate model sensitivities. The same authors performed similar studies to address the problem of correlation among model inputs (Ji and Prevedouros, 2005b), the a priori knowledge of model input probability distribution (Ji and Prevedouros, 2006), and a comparison with several other techniques not reviewed here (Ji and Prevedouros, 2007).

5.3.7 Monte Carlo filtering

When one is not interested in studying the specific value of Y and Y is above or below a certain threshold (if Y creates or does not create a certain effect), Monte Carlo filtering can be used. By using a Monte Carlo setting to produce the matrices and vectors of Equations (5.2) and (5.3) and then applying the filter of interest to the values of Y, it is possible to divide the matrix M of Equation (5.2) in two groups—one for the variable values producing one effect and the other for those that do not produce the effect. At this point, a statistical test can be conducted to check whether each input is statistically responsible for the effect to be produced.

No applications of this technique to the field of transportation modeling can be found. However, this technique is commonly used in other research fields (see Saltelli et al., 2005).

5.3.8 Metamodeling

A possible way to perform sensitivity analysis of complex black-box models is to use a metamodel that can approximate the output of the model itself. In this way, the time required is used to create the metamodel and the analysis can be performed easily using its analytical formulation. This topic is attracting the interest of researchers, in particular because of the interesting properties of some of the metamodels. An interested reader can refer to Chapter 5 of Saltelli et al. (2008).

An application of this technique to compare different goodness-of-fit functions for the calibration of the AIMSUM software can be found in Ciuffo et al. (2010).

5.3.9 Analysis of variance (ANOVA)

The ANOVA is a model-independent probabilistic sensitivity analysis method used to determine a statistical association between an output and one or more inputs. ANOVA differs from regression analysis in that no assumption is needed regarding the functional forms of relationships of inputs and outputs.

In ANOVA, model inputs are referred to as factors and their values are called factor levels. An output is designated a response variable. Multifactor ANOVA studies the effects of two or more factors on a response variable, and it is used to determine both first-order and interaction effects between factors and the response variable.

To apply this technique, a number of evaluations of the responses against different values of the input parameters are required. From a statistical view, an appropriate way of performing these evaluations is defined by the experimental design techniques. In particular, a full factorial design can be properly applied in this case so that the full factorial experimental plan consists of $n \cdot k$ model evaluations, where k is the number of factors and n is the number of levels.

The use of this experimental plan can also determine whether the factors interact with each other, that is, whether the effect of one factor depends on the levels of the others. The results that can be obtained from the ANOVA are twofold. First, they give an estimation of the model output variance explained by each parameter or by their combination. On the basis of this result, it is then possible to use a Fisher probability distribution to test the null hypothesis that the variance explained by a single parameter is negligible with respect to the whole model, that is, that the model is not sensitive (with a well-defined level of significance) to parameter changes.

For further details on experimental design techniques and on the ANOVA implementation, the interested reader could refer to technical books such as Law and Kelton (2000) or Box et al. (1978).

Despite the widespread use of ANOVA among the experimental disciplines and in the world of simulation, few studies to date have performed ANOVA to ascertain the impacts of the parameters of traffic flow models. Indeed, even a more general model sensitivity analysis is not commonly found in the literature, and, when it does appear, the analysis usually plays only a marginal role in the overall model description (Schultz et al., 2004; Jayakrishnan et al., 2001).

An exception is the sensitivity analysis carried out on the PARAMICS model by Bartin et al. (2006) and Li et al. (2009) to draw inference about the first-order effects of traffic model parameters. The interaction effects were not captured since a two-level full factorial design was adopted. A three-level factorial design was used instead in Beegala et al. (2005), in Ciuffo et al. (2007), and in Punzo and Ciuffo (2009) for the AIMSUN model. However, second-order interactions effects of parameters could be evaluated only in the last two studies in which a full factorial design was adopted—unlike Beegala et al. who adopted a fractional design.

Another study that used ANOVA to undertake a model sensitivity analysis was Park and Qi (2005). They used ANOVA to select a parameter from a set of eight parameters that needed calibration. Five levels per parameter were taken into account. A Latin hypercube sampling algorithm was used to define the experimental design. However, the interaction effect of the parameters was not evaluated.

5.3.10 Fourier amplitude sensitivity test (FAST)

FAST is a method based on Fourier transformation of uncertain model parameters into a frequency domain, thus reducing the multidimensional model into a single-dimensional one (Frey and Patil, 2002; Cukier et al., 1978; Morgan et al., 1990; Saltelli et al., 1998 and 1999; Isukapalli, 1999). The FAST method can be used for both uncertainty analysis and sensitivity analysis and does not depend on assumptions about the model structure. FAST allows the computation of the fraction of the variance of a given model output attributable to each input parameter.

Consider the function $y = f(x)$, which is defined by a set of n input parameters in the n-dimensional unit cube (R^n):

$$R^n = (x | 0 \leq x_i \leq 1; i = 1,\ldots,n) \tag{5.11}$$

The essence of this technique is to generate a curve in the parameter space that is a periodic function of each parameter, with a different frequency for each variable. The curve is defined by a set of parametric equations. The one proposed by Saltelli et al. (1999) is shown below and is used in several applications:

$$x_i(s) = \frac{1}{2} + \frac{1}{\pi} \arcsin(\sin \omega_i s) \tag{5.12}$$

where s is a scalar variable that varies over the range $-\infty < s < \infty$, and $\{\omega_i\}$, $\forall i = 1,2,\ldots,n$, is a set of different frequencies associated with each factor. As s varies, all the factors change simultaneously along a curve that explores R^n. Each x^i oscillates periodically at the corresponding frequency ω_i. The output y shows different periodicities combined with the different frequencies ω_i for any model f. If the i-th factor has a strong influence on the output, the oscillations of y at frequency ω_i have high amplitudes.

This is the basis for computing a sensitivity measure that for factor x_i is based on the coefficients of the corresponding frequency ω_i and its harmonics. The original function $y = f(x)$ is transformed into $y = f(s)$ when each input parameter is represented by Equation (5.12). The function $y = f(s)$ can be expanded into a Fourier series:

$$y = \sum_{j=-\infty}^{j=+\infty} (A_j \cos js + B_j \sin js) \tag{5.13}$$

where A_j and B_j are two common Fourier coefficients of the cosine and the sine series, respectively. If the spectrum of the Fourier series expansion is defined as $\Lambda_j = A_j^2 + B_j^2$ with $j \in Z = \{-\infty,\ldots,0,\ldots,\infty\}$, the total variance of the output $(\hat{D})$ is:

$$\hat{D} = \sum_{j \in Z} \Lambda_j \tag{5.14}$$

The portion of the output variance contributed by the input parameter i $(\hat{D}_i)$ is defined as:

$$\hat{D}_i = \sum_{j \in Z_0} \Lambda_{p\omega_i} \tag{5.15}$$

where Z_0 is equal to $Z - (0)$, and p is an integer. The importance of each parameter for the uncertainty of the output can be measured by the percentage of the contributed variance.

In the field of traffic modeling and simulation, the FAST method has been applied only by Ji and Prevedouros (2005a) to compare the performances of different sensitivity analysis techniques to the investigation of the delay model proposed in the 2000 edition of the HCM.

5.3.11 Variance-based methods on Sobol decomposition of variance

The variance-based method based on the Sobol decomposition of variance has been left to the end of the list since it is one of the most recent and effective global sensitivity analysis techniques. For this reason, it deserves detailed discussion. The original formulation of the method is due to Sobol (1993 and 2001) who provided the analytical derivation and the Monte Carlo–based implementation of the concept. The latest setting for its practical implementation is from Saltelli et al. (2009). A brief description of the method follows.

Let us consider again the model of Equation (5.1). We want to see what happens to the uncertainty of Y if we fix one of the input variables Z_i to a specific value z_i^*. The resulting variance is, in general, lower than the unconditional variance of y will be $V_{Z_{-i}}(Y|Z_i = z_i^*)$. The symbolism in Z_{-i} means that we are considering the variance across all the variables but the i-th. The larger the influence of variable Z_i, the lower will be the proportion between the conditional variance and the total variance of Y. For this reason, the conditional variance can be considered an index of sensitivity for Z_i. The problem with this formulation is that the sensitivity index depends on the specific value z_i^* considered. Therefore, we consider the average of this measure over all possible points z_i^*, $E_{Z_i}(V_{Z_{-i}}(Y|Z_i))$. Furthermore, it is known that

$$V(Y) = E_{Z_i}(V_{Z_{-i}}(Y|Z_i)) + V_{Z_i}(E_{Z_{-i}}(Y|Z_i)) \tag{5.16}$$

Equation (5.16) shows that for Z_i to be an important factor, $E_{Z_i}(V_{Z_{-i}}(Y|Z_i))$ must be small, that is, the closer $V_{Z_i}(E_{Z_{-i}}(Y|Z_i))$ is to the unconditional variance $V(Y)$, the greater the influence of Z_i. Thus, we may define our first-order sensitivity index of Z_i with respect to Y as:

$$S_i = \frac{V_{Z_i}(E_{Z_{-i}}(Y|Z_i))}{V(Y)} \tag{5.17}$$

For a comprehensive physical interpretation of such index, refer again to Saltelli et al. (2008).

The first-order sensitivity index is a very important measure for determining how much the correct definition of a model input may reduce the overall variance of results. From Equations (5.16) and (5.17), we have $S_i \leq 1$. It is possible to define a model as additive if:

$$\sum_{i=1}^{r} S_i = 1 \tag{5.18}$$

In this case, the unconditional variance of the model can be decomposed in the sum of the first-order effect of each single variable. Usually this is not the case, meaning that the joint combination of some variables can be responsible for a certain share of the unconditional variance, that is, just the definition of nonadditive models. In this case, a low first-order sensitivity index does not necessarily imply that the corresponding variable has scarce effect on the output variance, since it may contribute considerably to the total output variance via its combination with the other variables.

For this reason, using the so-called ANOVA-HDMR (analysis of variance–high dimensional model representation) decomposition developed by Sobol (1993) makes it possible to say that a full analysis of a model with r variables requires all the elements of the following equation to be discovered (in number of $(2^r - 1)$):

$$\sum_{i=1}^{r} S_i + \sum_{i=1}^{r} \sum_{j>i} S_{i,j} + \sum_{i=1}^{r} \sum_{j>i} \sum_{l>j} S_{i,j,l} + \cdots + S_{1,2,3,\ldots,r} = 1 \tag{5.19}$$

However, the characterization and practical evaluation of all the sensitivity indices in Equation (5.19) would be an expensive undertaking. To reduce the efforts required, a synthetic indicator to be coupled with the first-order sensitivity index constitutes the total effects index, defined as follows (Homma and Saltelli, 1996; Saltelli, 2002):

$$S_{T_i} = 1 - \frac{V_{Z_{\sim i}}(E_{Z_i}(Y|Z_{\sim i}))}{V(Y)} = \frac{E_{Z_{\sim i}}(V_{Z_i}(Y|Z_{\sim i}))}{V(Y)} \tag{5.20}$$

The total effects index of the input factor i provides the sum of all the elements in Equation (5.19), including the i-th. When the total index is $S_{T_i} = 0$, the i-th factor can be fixed without affecting the outputs' variance. If $S_{T_i} \cong 0$, the approximation made depends on the value of S_{T_i} (Sobol et al., 2007). It is worth noting that while $\sum_{i=1}^{r} S_i \leq 1$, $\sum_{i=1}^{r} S_{T_i} \geq 1$, both are equal to one only for additive models.

Since the analytical feasibility of traffic flow models limits the use of the formulas for the calculation of the variances reported in Equation (5.16), this method can be performed effectively in a Monte Carlo setting. In the field of traffic flow modeling, to the best of the authors' knowledge, this method based on quasi–Monte Carlo sampling in the parameter space has been used only by Punzo and Ciuffo (2011) to individuate the subset of parameters to calibrate in the case of two well-known car-following models. However, no other studies in this context made use of this technique to perform global sensitivity analysis.

5.3.12 Variance-based methods: implementation

In the previous section, we presented the first-order and total effects sensitivity indices that will be used in our application. Their simultaneous evaluation, however, is not straightforward. In the present work, we use the methodology described in Saltelli et al. (2008 and 2010) and presented hereafter.

A simpler way to evaluate sensitivity indices is the computation of the multidimensional integrals (in a Monte Carlo setting) in the space of the input variables. This would require N^2 model evaluations for each sensitivity index; N again is the size of the Monte Carlo experiment.

This has been revealed to be not necessary (Saltelli, 2002) and the following procedure was adopted:

First, two (N,r) matrices of quasi-random numbers (Sobol, 1976) are generated. Using the random numbers, two matrices of values (A and B) for the input variables of the model as in Equation (5.1) are generated. The tool for quasi-random number generation is freely available on the Internet (Simlab, 2010).

$$A = \begin{bmatrix} z_1^{(1)} & z_2^{(1)} & \cdots & z_r^{(1)} \\ z_1^{(2)} & z_2^{(2)} & \cdots & z_r^{(2)} \\ \cdots & \cdots & \cdots & \cdots \\ z_1^{(N)} & z_2^{(N)} & \cdots & z_r^{(N)} \end{bmatrix} \tag{5.21}$$

$$B = \begin{bmatrix} z_{r+1}^{(1)} & z_{r+2}^{(1)} & \cdots & z_{2r}^{(1)} \\ z_{r+1}^{(2)} & z_{r+2}^{(2)} & \cdots & z_{2r}^{(2)} \\ \cdots & \cdots & \cdots & \cdots \\ z_{r+1}^{(N)} & z_{r+2}^{(N)} & \cdots & z_{2r}^{(N)} \end{bmatrix} \tag{5.22}$$

Second, a set of r matrices designated C is obtained by assembling r matrices equal to A except for the i-th column (with i varying from 1 to r among the r matrices) taken from B.

$$C_i = \begin{bmatrix} z_1^{(1)} & z_2^{(1)} & \cdots & z_{r+i}^{(1)} & \cdots & z_r^{(1)} \\ z_1^{(2)} & z_2^{(2)} & \cdots & z_{r+i}^{(2)} & \cdots & z_r^{(2)} \\ \cdots & \cdots & \cdots & \cdots & \cdots & \cdots \\ z_1^{(N)} & z_2^{(N)} & \cdots & z_{r+i}^{(N)} & \cdots & z_r^{(N)} \end{bmatrix} \quad for \;\; i=1...r \tag{5.23}$$

Third, the model is evaluated for all $[N \cdot (r+2)]$ combinations of input variables as given by matrices A, B, and C to produce the vectors of output $y_A = f(A)$, $y_B = f(B)$ and $y_{C_i} = f(C_i)$ for $i = 1,...,r$. These vectors are sufficient for the evaluation of all the first-order and total effects indices. This is why the application of this technique for variance-based methods requires $[N \cdot (r+2)]$, which is still a not negligible number for complex and expensive models but is definitely lower than $N^2 \cdot r$.

Sensitivity indices can be then evaluated using the following formulations (Saltelli et al., 2010):

$$S_i = \frac{\frac{1}{N}\sum_{j=1}^{N} y_B^{(j)}\left(y_{C_i}^{(j)} - y_A^{(j)}\right)}{\frac{1}{2N}\sum_{j=1}^{N}\left(y_{A+B}^{(j)}\right)^2 - \left(\frac{1}{2N}\sum_{j=1}^{N} y_{A+B}^{(j)}\right)^2} \tag{5.24}$$

$$S_{T_i} = \frac{\frac{1}{2N}\sum_{j=1}^{N}\left(y_A^{(j)} - y_{C_i}^{(j)}\right)^2}{\frac{1}{2N}\sum_{j=1}^{N}\left(y_{A+B}^{(j)}\right)^2 - \left(\frac{1}{2N}\sum_{j=1}^{N} y_{A+B}^{(j)}\right)^2} \tag{5.25}$$

The choices of N and the input variable distribution are the last points to be discussed in this section. There are no universal recipes for these cases. N can vary from a few hundred to several thousand. A possible strategy to be adopted is to evaluate the indices per $\hat{N}$ in the range $[1, N]$, with N a "sufficiently" large number. By plotting the sensitivity indices against $\hat{N}$, it is possible to know whether they reached a stable value (i.e., a value no longer depending on N). If they are not stable, it is possible to increase again N. For this reason, the results of the sensitivity analysis will be shown in graphical form.

For the distribution of the input variables, the only possibility is relying upon a priori information (physical meanings of the variables, previous studies, expert suggestions, etc.). If such information is not available, preliminary tests should be performed to find the best settings.

5.3.13 Further comments

A final issue concerns correlated inputs. No simple or computationally efficient solution may exist. For alternative strategies, we suggest the reader refer to Saltelli and Tarantola (2002), Jacques et al. (2006), and Da Veiga et al. (2009).

5.4 CONCLUSIONS AND RECOMMENDATIONS

Appendix B at the end of this book presents a complete description of variance-based technique application to sensitivity analysis studies for two car-following models. The reader is referred to the appendix for an illustration of the techniques used.

This chapter has presented a broad overview of the various sensitivity analysis methods found in the literature. We encourage the use of such methods that take into account the combined effects of various types of parameters.

Chapter 6

Network model calibration studies

Tomer Toledo, Tanya Kolechkina, Peter Wagner, Biagio Ciuffo, Carlos Azevedo, Vittorio Marzano, and Gunnar Flötteröd

CONTENTS

This chapter details how calibration and estimation work in the realm of large networks. The chapter is mainly related to OD estimation, which is arguably the most important step before applying a simulation model (whether microscopic or macroscopic) to a certain study question. Section 6.1 deals with the calibration of network models in general. Section 6.2 covers demand models, followed by details on supply models in Section 6.3.

Section 6.4 presents a joint calibration of demand and supply models. The chapter ends with a summary in Section 6.5.

6.1 CALIBRATION OF NETWORK MODELS

Calibration variables associated with a simulation model are grouped based on the constituent model components whose inputs and parameters are under consideration. In general, simulation model components are divided into two categories: (1) demand models that estimate and predict OD trip flows and simulate travel behavior parameters; and (2) supply models that capture traffic dynamics. Hence, the problem of simulation model calibration involves estimation and calibration of all its components.

Generally, demand and supply models are calibrated separately. However, methodologies for joint demand–supply calibration have also been developed recently. The methods for sequential and simultaneous calibration of the demand and supply parameters using aggregate data are reviewed in the following sections.

6.2 DEMAND MODELS

Demand parameter calibration involves estimation of travel behaviors and OD demand flows. Travel behavior parameters are presented by pre-trip departure time, mode and route choice, and en route response to information. These parameters are generally estimated using disaggregate data that are not considered in this chapter. This class of parameters can be calibrated simultaneously with OD demand flows using aggregate data. The review of demand model estimation techniques begins with a discussion of OD demand matrix estimation and continues with presentation of methods for joint estimation of OD demand flows and travel behavior parameters.

6.2.1 OD demand estimation

Many approaches have been proposed to solve the OD estimation problem. They can be grouped into two main classes: direct estimation methods (surveys) and estimation methods using traffic counts. In general, the direct estimation methods are expensive and time consuming. In recent years, increasing attention has been devoted to more effective methods of OD demand estimation using traffic counts.

OD estimation methods based on traffic counts can be classified based on network performances and temporal demand dynamics. With respect to network performances, a distinction is made between congested and uncongested networks. Estimated travel demands are also classified as steady state (static) or time dependent (dynamic) matrices.

6.2.2 Static OD demand estimation methods

6.2.2.1 Uncongested networks

OD demand estimation methods for uncongested networks assume that link and path travel times can be computed from the network model even though link flows are not available. The methods proposed in the literature for the static OD demand estimation problem on uncongested networks include information minimization (Van Zuylen, 1979); entropy maximization (Van Zuylen and Willumsen, 1980; Brenninger-Gothe et al., 1989; Lam and Lo, 1991); maximum likelihood estimation, or MLE (Spiess, 1987; Cascetta and Nguyen, 1988; Lo et al., 1996; Hazelton, 2000); generalized least squares, or GLS (Cascetta, 1984; Bell, 1991); and Bayesian inference (Maher, 1983).

Following Cascetta and Nguyen (1988) and Cascetta (2001), classical estimators provide for the MLE of a demand vector by maximizing the probability (likelihood) of observing both OD sampling survey data and link counts (under the usually acceptable assumption that these two probabilities are independent), yielding:

$$d_{\mathrm{ML}} = \arg\max_{x \geq 0}[\ln\, L(\hat{d}|x) + \ln\, L(\hat{f}|x)] \tag{6.1}$$

where x is the variable demand, $\hat{d}$ is the demand by sample, and $\hat{f}$ is the vector of link counts. The log likelihood functions in Equation (6.1) are based on hypotheses on the probability distribution of demand counts $\hat{d}$ and traffic counts $\hat{f}$, respectively, conditional on the demand vector x.

Normally, traffic counts can be assumed to be independently distributed as Poisson random variables or follow a multivariate normal random variable, while the statistical distribution of OD demand counts depends on the sampling strategy. Generalized least squares (GLS) demand estimation d_{GLS} provides an estimate of the OD demand flow, starting from a system of linear stochastic equations, leading to the following optimization problem:

$$d_{\mathrm{GLS}} = \arg\min_{x \geq 0}\left\{\frac{1}{2}(x - \hat{d})^T Z^{-1}(x - \hat{d}) + \frac{1}{2}(\hat{f} - M_f x)^T W^{-1}(\hat{f} - M_f x)\right\} \tag{6.2}$$

where M_f is the submatrix of the assignment matrix related to links with available traffic counts and Z and W the covariance matrices related to the sampling error underlying the demand estimation and the measurement or assignment errors, respectively.

Bayesian methods estimate unknown parameters by combining experimental information (traffic counts in this case) with nonexperimental information (a priori or "subjective" expectations on OD demand, e.g., from an out-of-date estimation or a model system) by maximizing the logarithm of a posteriori probability:

$$d_B = \arg\max_{x \geq 0} [\ln g(x|d^*) + \ln L(\hat{f}|x)] \tag{6.3}$$

where $g(x|d^*)$ expresses the distribution of subjective probability attributed to the unknown vector given the a priori estimate d^* and $L(\hat{f}|x)$ expresses the probability of observing the vector of traffic counts $\hat{f}$ conditional on the unknown demand vector x. Again, the detailed specification of a Bayesian estimator depends on the assumptions about the probability functions $g(x|d^*)$ and $L(\hat{f}|x)$.

Normally, the unknown demand vector can be assumed to follow a multinomial random variable (in this case $\ln g(x|d^*)$ becomes the entropy function of the unknown vector x), a Poisson random variable (in this case $\ln g(x|d^*)$ becomes the information function of the unknown vector x), or a multivariate normal random variable. Note that approaches (6.1) through (6.3) yield objective functions of identical form if all the relevant distributions are assumed to be multivariate normal.

6.2.2.2 Congested networks

In a congested network, travel times, path choice fractions, and assignment fractions depend on link traffic flows. Generally, neither travel times nor flows are known on all links. Instead, they are computed by applying an assignment model, the result of which in turn depends on the assigned OD matrix. The problem of the circular dependence between OD matrix estimation and traffic flow assignment on congested networks has been studied by a number of authors.

A bilevel programming approach is one of the possible solutions to ensure the interdependency of OD demand matrix and traffic assignment. In the bilevel approach, the upper-level problem uses one of the statistical techniques proposed earlier (e.g., maximum entropy, GLS, MLE) to select the most appropriate OD matrix. The lower-level problem, using deterministic or stochastic user equilibrium models (Sheffi, 1985), endogenously determines route choice proportions that are compatible with the estimated OD flows.

Several methods were proposed to handle the bilevel program. Fisk (1988) combined the entropy maximum model with equilibrium conditions to construct a bilevel programming problem and used a variational inequality formulation to solve it. Heuristic iterative algorithms for the bilevel estimation problem solution were proposed by Spiess (1990),

Yang et al. (1992), Yang (1995), Yang and Yagar (1995), and Maher and Zhang (1999). The models proposed by Cascetta and Postorino (2001), Maher et al. (2001), and Yang et al. (2001) differ from the previous models in that they assume a stochastic user equilibrium.

Florian and Chen (1992 and 1995) reformulated the bilevel problem into a single-level equilibrium-based OD demand estimation problem using the concept of marginal function. However, most of these methods are acceptable only for small-and medium-scale networks. Lundgren and Peterson (2008) extended the earlier proposed methods (Spiess, 1990; Maher and Zhang, 1999) to adapt them for large-scale networks. Kim et al. (2001) and Al-Battaineh and Kaysi (2006) used genetic algorithms as an alternative approach for solution of the bilevel OD estimation problem.

Cascetta and Postorino (2001) formulated the general static OD estimation problem for congested networks as a fixed-point problem. To solve the fixed-point problem and obtain consistent OD flows, iterative schemes based on the method of successive averages (MSA) were applied. The sequential GLS-based OD estimator was used to generate updated flows in each iteration.

Lo et al. (1996) introduced an explicit representation of the stochastic nature of observed flows, eventually generalized by Vardi (1996). Lo et al. (1999) described an optimization method for applying this approach to large-scale networks. Hazelton (2000) proposed a method that makes use only of link counts but requires explicit path enumeration and is in practice extremely time-consuming for large networks. Finally, as noted by Hazelton (2003), a promising research development entails considering time series link counts (several days) as a key aspect for improving the reliability of OD matrix estimation. He took the covariance matrices of link count observations over several days into the estimation procedure, showing its reliability on very small test networks.

6.2.3 Dynamic OD demand estimation methods

Static OD estimation models calculate a single matrix of mean OD demand values using the average traffic counts collected over a relatively long period. The time-varying nature of link flows and OD demands is ignored in these matrices, disregarding the departure time offsets between vehicles counted at some link and affecting their applicability in dynamic traffic management problems.

In recent years, different dynamic OD demand estimation models have been proposed to overcome the limitations of the static models. Dynamic methods can be categorized into on- and off-line approaches. Off-line methods simultaneously estimate demand and supply model parameters. The on-line methods jointly update in real time the demand and supply parameter values estimated during the off-line step to better reflect prevailing conditions (Antoniou et al., 2009).

6.2.3.1 On-line estimation methods

Okutani (1987) developed the first on-line OD estimation model suitable for general networks. An autoregressive model is used to describe the evolution of OD flows over time. A Kalman filter was used to obtain optimal estimates of the state vector in each time interval. However, no information was provided on how the dynamic assignment fractions can be determined. Furthermore, the autoregressive process may not capture transitions of time-dependent demand patterns such as patterns from peak to off-peak periods.

Ashok and Ben-Akiva (1993) presented a Kalman filter–based approach for on-line estimation and prediction. To improve Okutani's model, they introduced deviations of OD demand flows from historical estimates. The proposed model was expected to estimate and predict the within-day deviations of travel demands from their average day-to-day pattern. This model overcomes deficiencies of several predecessors but has some shortcomings. First, no attempt is made to capture errors in the assignment matrix. Second, the model requires augmenting the state for a given interval with states corresponding to several prior intervals. This increases the computational load, thereby making an on-line application of the model difficult.

In subsequent work, Ashok (1996) indicated that the assignment matrix should be estimated because it is computed from random variables such as link travel times. Two methods were suggested to obtain an assignment matrix. The first approach involves the use of a traffic simulator to load the best current OD flows onto the network. The assignment matrix then can be calculated through analysis of vehicle records at sensors. Another approach is computing the assignment fractions for a given sensor and OD pair by summing the product of crossing fractions and the corresponding path choice fractions across all paths.

Ashok and Ben-Akiva (2000) suggested an alternative approach, having redefined the state variables as the deviations of departure flows from each origin and the shares headed to each destination. Except for different forms of transition equations, the approach has a similar framework as those they proposed earlier.

Ashok and Ben-Akiva (1993 and 2000) reported encouraging results in their case studies and field tests, and indicated that their methods are robust. In the algorithm proposed by Bierlaire and Crittin (2004), the state described by Ashok and Ben-Akiva is combined with a general technique for solving large-scale least squares problems. This iterative algorithm is more efficient than the Kalman filter approach suggested by Ashok and Ben-Akiva.

6.2.3.2 Off-line estimation methods

The off-line dynamic OD estimation problem was first formulated by Cascetta et al. (1993). The authors generalized the statistical framework proposed for the static problem and extended it to the dynamic OD estimation case. Estimates of dynamic OD demand flows are obtained by optimizing a two-part objective function. The first part measures the difference between the estimated OD matrix for an interval and the historic estimate of the OD matrix for that interval. The second part measures the difference between measured link volumes and those predicted by the model when the estimated OD demand flows are assigned to the network.

The authors used the GLS method, combining traffic counts with other available information on OD flows such as earlier matrices and surveys. Two types of estimators (simultaneous and sequential) were proposed. The simultaneous estimator designed for off-line applications yields in one step the entire set of time-dependent OD demand vectors by using link traffic counts from all time intervals. The sequential estimator suitable for on-line or large-scale applications gives in each step the OD vector for a given time interval by using both previous OD estimates and the current and previous traffic counts. The simultaneous estimator, while statistically more efficient than its sequential approximation, entails significant computational overhead that precludes its application to large problems.

Cascetta et al. noted that link travel times can be obtained from a traffic surveillance system or from a day-to-day dynamic traffic assignment model. Either way, the implication is that the dynamic assignment matrix is exogenously determined so that it might be inconsistent with the estimated assignment mapping if the network is congested. This model was further developed by Tavana (2001) and Ashok and Ben-Akiva (2002).

Tavana and Mahmassani (2000) and Tavana (2001) proposed a bilevel optimization model and an iterative solution framework to estimate dynamic OD demand matrix. In the upper-level problem, the demand value is estimated by minimizing the sum of squared errors in traffic counts with the OD matrix entries. This optimization problem was solved using the conventional GLS estimator and assuming that the link flow proportions were constant. The dynamic user optimal conditions were treated as constraints.

The resulting dynamic estimation problem was solved heuristically using an iterative optimization assignment algorithm. Specifically, the dynamic assignment factors were updated iteratively using the dynamic user equilibrium solution.

Tavana's model was extended by Zhou and Mahmassani (2003). They proposed a multiobjective optimization framework to combine available historical static demand information and multiday traffic link counts to estimate variations in traffic demand over multiple days.

Lindveld (2003) also proposed an iterative heuristic for the dynamic OD demand estimation problem. The problem was formulated as a bilevel optimization problem. The estimate of dynamic OD matrix is found on the upper level, using both traffic counts and a priori demand information, with dynamic traffic equilibrium constraints on the lower level.

The proposed model is basically a time-dependent extension of the static model proposed by Maher and Zhang (1999) for deterministic user equilibrium, and further developed by Maher et al. (2001) for stochastic user equilibrium. The method has been successfully implemented for a small test network with a corridor structure, i.e., a network with no route choices.

Sherali and Park (2001) proposed a constrained least squares (CLS) model but solved for path flows rather than OD demand flows. Following Cascetta et al. (1993), it was assumed that the dynamic assignment matrix can be obtained externally. The proposed path generation algorithm begins with solving a restricted basic problem based on an initial choice of a set of OD paths, then augments the basic problem iteratively with time-dependent shortest paths upon a time-space expanded network. The method will terminate and claim an optimum if no new time-dependent shortest paths can be found.

Most of the presented models were developed for small networks or they simplify the congestion effects on route choice and travel time (Cascetta, 1993; Sherali and Park, 2001). Lundgren et al. (2006) proposed a heuristic for estimation of dynamic OD demand matrices using traffic counts. Special interest in this work was given to the assignment matrix that depends on demand in the case of congestion. The proposed method is an extension of the method of Maher and Zhang (1999) for a deterministic static case and further developed by Maher et al. (2001) for a stochastic static case.

The authors used an iterative algorithm based on the steepest descent method to find a solution to the estimation problem. The directional derivatives of the assignment map for the current OD demand matrix are approximated with a difference quotient that describes how a change of the OD demand matrix will induce a change of the link flows. However, this method was tested only on a relative small network and its effectiveness must be checked on larger networks.

6.2.4 Quasi-dynamic OD demand estimation methods

One of the main problems in the OD demand estimation methods using traffic counts is the large imbalance between equations and unknowns. In other words, the equations available for the demand estimate (equal to the number of observed link flows) generally have fewer unknowns (equal to the number of OD flows).

This is obviously true in both the static and dynamic cases. In passing from the static to the dynamic case, both equations and unknowns increase linearly proportionally to the number of time slices t considered. This circumstance significantly influences the quality of the OD demand estimation, which, as demonstrated by Marzano et al. (2009) in laboratory experiments, is strictly related to the quality of the a priori OD estimation.

An effective method to obtain a good-quality OD estimation with a poor-quality a priori OD estimate is making some assumptions about the demand evolution within a day (i.e., between different t *values*) to reduce the number of unknowns in a within-day dynamic context. As an example, while the generation profile of each zone could be considered time varying among the different time slices, the distribution percentages among the different destination zones could be considered linked to territorial aspects that vary more slowly across a day.

On the basis of this theoretical consideration, the aim of the OD demand estimation in a within-day dynamic context could be more efficaciously the estimation of generation profiles from each zone and for each t, using as distribution percentages the average values in a time period T larger than t. In this way, the equation-to-unknown ratio can be pushed toward the desired "one" value and true generation profiles. Average distribution percentages can be estimated with good quality independently of the quality of the a priori OD estimates, as shown by Marzano et al. (2009).

6.2.5 Joint estimation of OD demand and travel behavior parameters

The representation of demand in traffic simulation models relies on estimates of OD demand, route choice model parameters, and network travel times to accurately model a network and its demand patterns. Since calibration of OD demand flows and route choices separately leads to biased OD flow estimates, more focus on methods for joint estimation of demand parameters has grown in recent years. However, the literature on joint OD estimation and parameter calibration is still limited.

Liu and Fricker (1996) proposed a two-step heuristic method for sequentially estimating OD flows and route choice parameters on uncongested networks. In the first step, the route choice parameters are fixed and OD flows are estimated by minimizing the difference between observed and modeled link flows. In the second step, the link flows obtained from the first step are used to calibrate the route choice parameters using a maximum likelihood method. Iterations are repeated until the first derivative of the likelihood value approaches zero. However, apart from ignoring congestion effects, implementation of the model requires a complete set of link traffic counts.

Yang et al. (2001a) proposed an optimization model for simultaneous estimation of OD flows and a travel cost coefficient for congested networks in a logit-based stochastic user equilibrium model. The model was formulated in the form of a standard differentiable, nonlinear optimization problem with analytical stochastic user equilibrium constraints. Explicit expressions of the derivatives of the stochastic user equilibrium constraints with respect to OD demand, link flow, and travel cost coefficient were derived and computed efficiently through a stochastic network loading approach.

A successive quadratic programming algorithm using the derivative information was applied to solve the simultaneous estimation model. This work, however, performed only static OD estimates. Also, due to the inherent non-convex property of the problem, the technique used might lead to local optima.

Balakrishna (2002) presented an iterative framework for joint calibration of dynamic OD flows and route choice models using several days of traffic sensor data. The methodology for the calibration of the OD estimation module was based on an existing framework adapted to suit the sensor data usually collected from traffic networks. A static OD matrix for the morning peak was adjusted systematically using a sequential GLS estimator to obtain dynamic OD matrices for the entire peak. The calibrated parameters included route choice model parameters, time-varying OD flows, variance–covariance matrices associated with measurement errors, and a set of autoregressive matrices that captured the special and temporal interdependence of OD flows.

Sundaram (2002) developed a simulation-based short-term transportation planning framework for joint estimation of dynamic OD flows and network equilibrium travel times. While the coefficients of the route choice model are not estimated, a consistent set of OD flows and travel times are obtained by iterating between an OD estimation module and a day-to-day travel time updating model. Sundaram's approach operates in two steps. Travel times are established for a given set of dynamic OD demands. The resulting equilibrium travel time estimates are used to recalculate assignment matrices for OD estimation. Travel times may then be computed again based on the new OD estimates if convergence has not been attained.

He et al. (2001) proposed the MLE approach to estimate the parameters of dynamic OD demand and route choice simultaneously with consideration of dependencies among paths and links and flow propagation information by utilizing time-dependent traffic data and historical traffic information. Consistent estimates of dynamic OD flows and route choice probabilities were obtained using approximations of joint probability distribution functions of link traffic flows on a network. The proposed method can be applied using link flow data alone. It is also possible to estimate parameters with measurement errors and incomplete data, especially when traffic counts are available only on a few links in a network.

A further generalization is contained in Lo and Chan (2003), who proposed a procedure for the simultaneous estimation of the OD matrix and route choice dispersion parameter for congested networks.

6.2.6 Disaggregate demand estimation from aggregate data

The traditional approach of generating (prior) OD matrices via a four-step process is being replaced by activity-based travel demand models (Ortuzar and Willumsen, 2004; Bowman and Ben-Akiva, 1998). This development has been enabled by methodological progress on the behavioral modeling side, the increased availability of disaggregate data, and vast improvements in computational facilities.

A shortcoming of the current situation is that despite the disaggregate nature of activity-based travel demand models, dynamic traffic simulations are still based on aggregate OD matrices. Consequently, any OD matrix estimator is constrained to the adjustment of trip-making behavior, without access to the much richer behavioral information provided by a disaggregate activity-based travel demand model.

It is well known that OD matrices can be avoided by coupling the activity-based travel demand model directly to a traffic simulation (Nagel and Flötteröd, 2012). In this setting, it is still possible to use traffic counts for demand estimation, only now the disaggregate travel behavior in the activity-based model is adjusted from the counts directly, without intermediate aggregation into an OD matrix. This method is based on a Bayesian argument outlined below (Flötteröd et al., 2011).

A stochastic traffic assignment simulation generates realizations of network conditions and travel behavior that can be considered as draws from a prior distribution that reflects the analyst's knowledge of the modeled system. Traffic counts constitute additional measurements that can be linked through a likelihood function to the disaggregate travel behavior in the simulation (Flötteröd and Bierlaire, 2009). In a Bayesian framework, the demand estimation problem becomes the adjustment of the activity-based model so that it generates realizations of the posterior distribution of the travel behavior based on the measurements. The joint estimation of travel behavior and demand model parameters is possible along the same lines.

This approach (1) accounts for and exploits the complex behavioral constraints implemented in the activity-based model, (2) is applicable under the same technical premises as any OD matrix estimator, (3) is capable of estimating OD matrices as a special case of an activity-based model with very limited degrees of freedom, and (4) has vast computational advantages over usual OD matrix estimators in that it constitutes a one-step estimator that adjusts the travel demand within the iterative loop of the simulation. This results in an almost negligible computational overhead when compared a plain simulation.

Flötteröd et al. (2012) demonstrated the applicability of this method to a real-world scenario with tens thousands of network links and hundreds of thousands of simulated travelers. The approach is implemented in a freely available software tool that reflects its continuous development (Flötteröd, 2009; See http://home.abe.kth.se/~gunnar/cadyts.html).

6.2.7 Selection of link count locations

A theoretical issue strictly linked to the problem of OD matrix correction using traffic counts is the selection of methods for the optimal location of link count sections, that is, the identification of the set of link flows providing maximum information about the underlying OD matrix, given a budget constraint. The problem is formalized in the literature by defining a reliability measure of the OD matrix estimation based on traffic counts and through the consistent prop-osition of optimization techniques or heuristics for finding a set of link counts (under given constraints) that maximizes this reliability measure.

Yang et al. (1991) introduced the maximum possible relative error (MPRE) as a reliability measure and formulated a quadratic optimization problem under the constraints of equations expressing counted flows as a function of OD matrix entries in the hypothesis of error-free counts and assignment matrices. The same hypotheses were shared by Bierlaire (2002), who proposed theoretical measures related to the volume of the feasibility set of link flows. Based on the numerical difficulty of calculating the volume of a polytope in large dimensions, they also studied the total demand scale (TDS) measure, defined as the difference between the maximal and mini-mal total values of the demand (sum of OD matrix entries) consistent with counted flows calculable by two constrained linear programming problems.

Both Yang et al. (1991) and Bierlaire (2002) discuss the need for setting finitely valued reliability measures—an issue also addressed by Yang and Zhou (1998) by means of the constraints defined by the so-called OD cov-ering rule. In addition, Yang and Zhou (1998) propose three other rules for the optimal link count section locations, i.e., the maximal flow fraction, the maximal flow interception, and the link independence rules. Notably, the definition of upper bounds for the feasibility set of the link flows is theoretical speculation. From a practical view, proper limits can be defined based on socioeconomic characteristics of each traffic zone, e.g., number of generated trips not larger than the population in the morning peak hour. Furthermore, the quality of the method proposed by Yang and Zhou (1998) is strictly conditioned on the reliability of the prior OD matrix estimate.

A number of other methods have been proposed in the literature to date. Ehlert et al. (2006) generalized the work by Yang et al. (1991), introducing section-specific counting costs and count section weights into the optimi-zation functions related to their information content (or entropy). Another research path deals with identifying a whole set of link flows starting from an observed (counted) subset, as proposed by Hu et al. (2009) who utilize the sole topology of the network, requiring explicit path enumeration and without inferences about the underlying OD matrix.

Similarly, heuristics intended to provide a geographical or topological disaggregation of link flows are discussed by Yang et al. (2006). Other network-based flow measures can be taken into account as well, for exam-ple, the use of plate scanning surveys as proposed by Minguez et al. (2010), leading to the proposition of a mixed integer programming problem.

Simonelli et al. (2011) propose an innovative method for addressing opti-mal link count section location based on a reliability measure in which the prior accuracy of the OD matrix estimate, that is, its statistical distribution rather than its prior punctual estimate, is explicitly considered along with its posterior distribution conditioned on a specific subset of link count locations.

6.3 SUPPLY MODELS

In comparison with the demand parameters, the supply parameter calibration has attracted somewhat less attention even though it plays a critical role in determining network performance. Supply models mimic traffic dynamics using speed density relationships and the dissipation and spillback phenomena of traffic queue formation. The numbers and natures of supply variables may change, depending on the level of detail employed for capturing traffic dynamics and queuing phenomena. Recently developed algorithms applied to macroscopic, mesoscopic, and microscopic supply model calibration are reviewed in this section.

6.3.1 Macroscopic and mesoscopic supply calibration

Macroscopic traffic models capture traffic dynamics through aggregate relationships derived by approximating vehicular flow as a fluid. Several macroscopic models are reported in the literature, including METANET (Messmer and Papageorgiou, 2001), EMME/2 (INRO, 2006), VISUM (PTV, 2006), SATURN (Van Vliet, 2009), and the cell transmission model (CTM; Daganzo, 1994).

Mesoscopic models are syntheses of microscopic and macroscopic modeling concepts. They couple the detailed behavior of individual drivers' route choice behaviors with more macroscopic models of traffic dynamics. Examples of such systems include DynaMIT (Ben-Akiva et al., 2001 and 2002) and DYNASMART (Mahmassani, 2002).

In both mesoscopic and macroscopic traffic simulation models, speed density functions are critical for modeling traffic dynamics. Recent studies used systematic algorithms for supply model calibration, specifically speed density relationships, with varying degrees of success. The typical data used for the calibration of these parameters are sensor records of at least two of three primary traffic descriptors: speeds, flows, and densities. The applications are classified as off-line (archived sensor data) and on-line (real-time sensor data).

6.3.1.1 Off-line calibration approaches

Van Aerde and Rakha (1995) described the calibration of speed flow profiles by fitting data from loop detectors on Interstate 4 near Orlando, Florida. Network links were grouped based on the traffic characteristics observed at sensor locations. A speed flow profile estimated for a link equipped with a sensor was allotted among links in a group. Similar approaches have been widely applied on networks of realistic size and structure.

Leclercq (2005) estimated four parameters of a two-part flow-density function with data from arterial segments in Toulouse, France. The function contained a parabolic free flow part and a linear congested regime. An interior point conjugate gradient method was employed to optimize the fit to observed sensor flows, with the fitted flows obtained from the assumed flow density function. A major drawback of this approach is local fitting. The estimated link performance functions reflect spot measurements at discrete sensor stations and do not necessarily correspond to overall link dynamics (especially in congestion conditions). The estimation procedure also does not enforce consistency across contiguous links or segments. This indicates the need for an expanded approach that considers larger sections of networks.

Munoz et al. (2004) described a calibration methodology for a modified cell transmission model (MCTM) for a freeway stretch in California. Free flow speeds were obtained through least squares by fitting a speed flow plot through each detector's data. Free flow speeds for cells without detectors were computed by interpolating between the available speed estimates. In the case of poor or missing sensor data, a default of 60 mph was assumed. For the purpose of capacity estimation, the freeway was divided into congested and free flow sections by studying speed and density contours from detector data.

Capacities in the free flow cells were set slightly higher than the maximum flow observed at the nearest detector. Bottleneck capacities were estimated to match the observed mainline and ramp flows just upstream of the free flow part of the bottleneck. Speed flow functions were obtained through constrained least squares on sensor data from congested cells.

Yue and Yu (2000) calibrated the EMME/2 and QRS II (Horowitz, 2000) models for a small urban network in the United States While no systematic calibration approach was outlined, the authors adjusted the free flow travel times and turning fractions to match detector count data. Such ad hoc procedures are unlikely to perform satisfactorily when applied to large-scale models and networks.

Ngoduy and Hoogendoorn (2003) calibrated METANET parameters using the Nelder-Mead method (a gradient-free algorithm working directly with objective function evaluations) to calibrate a freeway section in the Netherlands. The calibrated terms included fundamental diagram parameters such as free flow speed, minimum speeds, maximum density, and coefficients that captured the effects of merging, weaving, and lane drops.

Park et al. (2006) applied DynaMIT to a network in Hampton Roads, Virginia, estimated speed density functions for segments. They adopted the procedure of Van Aerde and Rakha (1995) and concluded that the initial calibration results needed adjustments to improve DynaMIT's overall ability to estimate and predict traffic conditions.

Kunde (2002) presented a calibration of the supply models within a mesoscopic simulation system. A three-stage approach to supply calibration was outlined. It utilized an increasing order of complexity (single segment, subnetwork, entire network) applied to a large-size mixed network in Irvine, California, using DynaMIT. The results of the network-wide calibration showed that the simultaneous perturbation stochastic approximation (SPSA) algorithm provided results comparable to those of the Box complex algorithm by using far fewer function evaluations, thus requiring much less run time.

6.3.1.2 On-line calibration approaches

Van Arem and Van der Vliet (1992) developed an on-line procedure to estimate capacity at a motorway cross section. The method was based on two assumptions. The first was the existence of a "current" fundamental diagram that depended on prevailing conditions. A method for establishing such fundamental diagrams based on on-line measurements of flow, occupancy, and speed was presented. The second assumption was that capacity could be estimated using this fundamental diagram and the notion of "maximum" occupancy. The capacity was estimated by substituting the current maximum occupancy into the current fundamental diagram.

Tavana and Mahmassani (2000) used transfer function methods (bivariate time series models) to estimate dynamic speed density relations from typical detector data. The parameters were estimated using past speed density data. The method was based on time series analysis, using density as a leading indicator. The resulting model for estimating speed and predicting its value for future time intervals was descriptive rather than behavioral.

Huynh et al. (2002) extended the work of Tavana and Mahmassani (2000) by incorporating the transfer function model into a simulation-based dynamic traffic assignment (DTA) framework known as DYNASMART. Furthermore, the estimation of speeds using the transfer function model was implemented as an adaptive process to a small mixed network in San Antonio, Texas, where the model parameters were updated on-line based on prevailing traffic conditions. A nonlinear least squares optimization procedure was incorporated into the DTA system to enable the estimation of the transfer function model parameters on-line. Results from simulation-based experiments confirmed that the adaptive model outperformed the nonadaptive model. The scope of this study, however, was limited to updating speeds on a single link using synthetic data. Qin and Mahmassani (2004) addressed the shortcomings of the approach by evaluating the same model with actual sensor data.

Wang and Papageorgiou (2004) presented a general approach to the real-time estimation of the complete traffic state along freeway stretches. They used a stochastic macroscopic traffic flow model and formulated it as a state space model they solved using an extended Kalman filter. This formulation allowed dynamic tracking of time-varying model parameters by including them as state variables to be estimated. A random walk was used for the transition equations of the model parameters. Wang et al. (2007) presented an extended application of this approach.

6.3.2 Microscopic supply calibration

Microscopic models capture traffic dynamics through detailed representations of individual drivers and vehicular interactions. Popular commercial microscopic software packages include CORSIM (FHWA, 2005), PARAMICS (Smith et al., 1995), AIMSUN2 (Barcelo and Casas, 2002), MITSIMLab (Yang and Koutsopoulos, 1996; Yang et al., 2000), VISSIM (PTV, 2006), DRACULA (Liu et. al, 1995), and Trans-Modeler (Caliper, 2006).

Microscopic calibration involves estimation of driving behavior parameters such as acceleration, lane changing, and car following. Microscopic data are complex to obtain and to calibrate. The difficulty of calibration problem arises because available data are usually aggregate measurements of traffic characteristics. Aggregate calibration is based on the interactions among all the components of a simulation model. Therefore, it is impossible to identify the effects of individual models on traffic flow when using aggregate data. In general, the aggregate calibration of microscopic supply parameters is a part of a joint demand–supply calibration. Furthermore, the calibration of microscopic models does not provide general results and use of the resulting set of parameters in applications in different locations is often difficult.

A number of published papers discussed microscopic calibration using aggregate data. Hourdakis et al. (2003) presented a three-stage general and systematic method for manually calibrating microscopic traffic simulators. They first sought to match observed traffic flows by calibrating global parameters such as vehicle characteristics, then local link-specific parameters such as speed limits were calibrated to match observed speeds. An optional third calibration stage was suggested by which any measure chosen by the user could be compared. A quasi-Newton algorithm was used for the solutions of various subproblems.

A simplex optimization approach for microscopic calibration was proposed by Kim and Rilett (2003). The simulated and observed traffic counts from TRANSIMS and CORSIM micro-simulation models were matched under two demand matrices. The gradient-free downhill simplex algorithm was used by Brockfeld et al. (2005) to calibrate a small set of supply parameters in a wide range of microscopic and macroscopic traffic models.

Hollander and Liu (2005) used a similar simplex approach for calibrating a small urban network in the UK with DRACULA by minimizing an objective function that expresses differences between observed and simulated travel time distributions.

The use of genetic algorithms (GAs) for microscopic parameter calibrations was illustrated in several works. Ma and Abdulhai (2002) developed a parametric optimization tool known as GENOSIM for micro-simulation calibration. In GENOSIM, GAs are used by a generic and independent calibration tool to interface with any microscopic traffic simulation model and the model calibration process is transformed into an automated, systematic, and robust process.

Kim and Rilett (2004) conducted calibration with GA by using both CORSIM and TRANSIMS and showed the benefits of using a GAs for automated calibration. Kim et al. (2005) proposed a nonparametric statistical technique for use in an automated microscopic calibration procedure. GA was used to find the best parameters. The proposed method was applied to a real network by using VISSIM.

These approaches used a few selected calibration parameters due to the complexity of the optimization surface (i.e., stochastic nature of microscopic simulation models and total number of combinations when all possible calibration parameters are considered). Park and Qi (2005) adopted a statistical experimental design approach to reduce the number of combinations and also considered the feasibility of the initial ranges of calibration parameters. The authors showed that if initial ranges of calibration parameters do not contain an optimal solution (i.e., field condition), the simulation model cannot be calibrated even if the optimization finds an optimal solution within the search region. This approach was successfully tested at an isolated signalized intersection using the VISSIM simulation model.

Hollander and Liu (2008) presented several guidelines after reviewing key aspects of the model calibration procedures cited in recent publications: scope of the calibration problem, formulation and automation of the calibration process (especially decomposition into subprocesses), measures of the goodness of fit, and number of repeated runs. They state that many calibration methodologies are not rigorous enough in terms of the number of repetitions of the model used throughout the calibration procedure and that most authors still use traffic micro-simulation for estimating mean values of various traffic measures even though the stochastic nature of micro-simulation creates an excellent opportunity for examining their variations.

Most optimization techniques applied for the calibration of supply model parameters are limited to simple networks and small parameter sets. To ascertain their suitability for overall model calibration, tests on larger networks and variable sets should be performed.

6.4 JOINT DEMAND–SUPPLY CALIBRATION

The calibration of demand and supply components has generally been attempted through a sequential procedure: supply parameters are calibrated first, assuming that demand inputs are known, then demand parameters are estimated with fixed supply parameters. In this case, complex interactions between demand and supply components are ignored, leading to suboptimal results. Instead, simultaneous estimation of demand-supply parameters captures these interactions, ensuring consistency among the estimated parameters. This section focuses on methods for joint demand–supply calibration.

Since the problem of joint calibration of demand and supply parameters has only recently arisen, literature on the topic is limited. Mahut et al. (2004) described the calibration of a mesoscopic traffic simulator. Their software combines a microscopic network loading model and a mesoscopic routing engine. Iterations between the two components are used to establish dynamic equilibrium travel times on the network.

In this calibration approach, the different model components are treated independently. OD flows are estimated by matching turning movement counts at major intersections with their simulated values. Capacities are estimated based on saturation flow rates and empirically derived downstream intersection capacities. Triangular volume-delay functions approximated based on posted speed limits and estimated capacities may not accurately reflect ground conditions since drivers on average travel at speeds higher than the speed limit under uncongested conditions. The gap acceptance and route choice parameters are adjusted manually to minimize the objective function. While valuable insights are provided to support these adjustments, they remain case-specific and may not be easily transferable.

Gupta (2005) demonstrated the calibration of the DynaMIT model by using separate methodologies to calibrate the demand and supply parameters. Supply and demand estimations were performed sequentially using real sensor count and speed data. Although the application has many limitations (such as sequential OD estimation and local supply fitting), it contributes significantly through the development of an observability test that allows the modeler to ascertain whether unique OD flows may be estimated from the given sensor configuration and coverage.

Several recent studies focused on calibrating both demand and supply model inputs for microscopic simulation. Mahanti (2004) calibrated the demand and select supply parameters for the MITSIMLab microscopic simulator by formulating the overall optimization problem in a GLS framework. The approach divides the parameter set into two groups: OD flows that may be estimated efficiently using existing tools and the remaining parameters (including a route choice coefficient, an acceleration–deceleration constant in the car-following model, and the mean and variance of the distribution of drivers' desired speeds relative to the speed limit). An iterative solution method is implemented, with the OD flows estimated using the classical GLS estimator and the parameters estimated by Box complex iterations.

Toledo et al. (2004) formulated the problem of jointly calibrating the OD flows, travel behavior, and driving behavior components of microscopic models using aggregate sensor data sources. The OD estimation step (utilizing a GLS formulation) was introduced as an explicit constraint, and a bilevel heuristic solution algorithm was used to solve for the three components iteratively. The use of the Box complex algorithm was reported for the estimation of select behavioral parameters.

Dowling et al. (2004) compared different simulated and observed measurements in separated stages of the calibration process: capacities to calibrate driving behavior parameters, flows to calibrate route choice parameters, and finally travel times and queue lengths to fine-tune all parameters. The root mean squared error was used as goodness of fit in the comparison of simulation outputs and observed measurements in a small number of locations of a mixed network.

Jha et al. (2004) calibrated MITSIMLab for a large-scale network. They estimated driving behavior parameters independently on a single freeway section for which OD flows could be inferred easily from sensor counts. Subsequently, OD flows, a route choice parameter, and habitual travel times were obtained by iteratively calibrating each component individually until convergence. The authors discussed several practical issues relevant to large-scale model calibration. The most important was the effect of stochasticity and extremely low OD flows on the quality of the simulated assignment matrices used for the GLS-based OD estimation.

Balakrishna et al. (2003) developed an off-line DTA model calibration method for simultaneous demand and supply parameter estimation. Two algorithms were used for optimization: SPSA (simultaneous perturbation stochastic approximation) and SNOBFIT. They concluded that the two algorithms produced comparable parameter estimates, although SPSA did so at a fraction of the computational requirements.

Balakrishna (2006) and Balakrishna et al. (2007a) presented a framework for simultaneous calibration of all supply and demand parameters typical of DTA models (OD flows, route choices, capacities, speed density parameters) using available data (counts, speeds, densities, queue lengths). The problem was solved with the SPSA algorithm. In the work proposed by Balakrishna et al. (2007b), the methodology was adapted for the simultaneous demand–supply calibration within a large-scale traffic microsimulation model.

Antoniou et al. (2007) formulated the problem of on-line calibration of a DTA model as a nonlinear state space model that allows the simultaneous calibration of all parameters and inputs. The methodology is generic and flexible, based on nonlinear extensions of the Kalman filter method:

the extended Kalman filter (EKF), the limiting EKF (LimEKF), and the unscented Kalman filter (UKF). The solution algorithms were applied to the on-line calibration of the state-of-the-art DynaMIT DTA model, and their use was demonstrated in a freeway network. The LimEKF showed accuracy comparable to that of the best algorithm but with vastly superior computational performance.

Vaze et al. (2009) presented a methodology for the joint calibration of demand and supply parameters of a DTA model using multiple data sources (traffic count and AVI data). The calibration problem was formulated as a stochastic optimization problem solved by using the GA and SPSA algorithms. The results indicated that use of AVI data significantly improved calibration accuracy.

6.5 SUMMARY

Modern traffic simulation models replicate various traffic phenomena using interactions among complex demand and supply components. To provide realistic abstractions of network processes, a large set of model inputs and parameters should be calibrated before a model is applied.

The goal of simulation model calibration is to obtain accurate depictions of certain traffic patterns: (1) dynamic OD demand matrices and route choice models to capture demand side effects; (2) detailed representation of network capacities and traffic dynamics to replicate network supply phenomena; and (3) complex interactions between travel demand and network supply. Simulation models should employ algorithms that can capture these interactions and accurately estimate queues, spillbacks, and delays.

This literature review indicates several shortcomings in the state of the art of simulation model calibration using aggregate data. Most approaches calibrate demand and supply components independently, ignoring the effects of their interactions. Supply parameters are calibrated first; then demand parameters are estimated with fixed supply data. Prevalent practices rely on heuristics and manual parameter adjustment approaches that are largely based on judgment. Moreover, these studies typically estimate a small subset of parameters deemed important in explaining observed data for specific networks and data sets, and typically do not perform sufficient iterations to ensure a high degree of accuracy.

In most of the existing methods for demand calibration, OD demand flows and route choice model parameters are estimated sequentially. OD demand estimates sometimes ignore congestion effects on route choices and travel times; or the fixed point problem of OD estimation is solved using bilevel approach with a fixed assignment matrix at each step of the optimization procedure.

With regard to supply parameter calibration, capacities are generally approximated from sensor flow data and from recommendations in the *Highway Capacity Manual*. Speed density functions are identified locally by fitting to sensor data. Most studies do not take into account the effects of various sources of variability in traffic counts and OD demand, such as variability in network conditions, events and incidents, weather, seasonality, and so on. A review of the existing methods for aggregate calibration can be concluded with the following main findings:

- Simultaneous calibration of all demand and supply parameters provides the most efficient estimates of a simulation model inputs.
- The most common technique for traffic data collection is using loop detectors for counts. Nevertheless, even though this tool can give time-dependent information about utilization at specific places, this type of data taken from various places does not produce sufficiently accurate information about vehicle utilization of a network in most applications. Using combined data from several tools leads to improved calibration accuracy.
- In comparison with manual search techniques, application of automated optimization algorithms (e.g., simplex algorithm) for calibration of simulation models is more efficient and less expensive. When a simulator is stochastic, the calibration problem must also account for the inherent noise in model outputs.
- Estimation of both travel times and OD demand flows in one process or estimation of OD flows using travel time information as additional measurement significantly improves estimation precision. Inclusion of travel time information into the calibration of the OD matrix allows a simulation model to better replicate the observed traffic conditions.
- Simultaneous estimation of time-dependent OD demand flows across multiple time intervals helps capture the effects of long trips. The traditional sequential method ignores the contributions of OD departure flows to measurements in future time intervals.
- Calibration of supply parameters is significantly improved by capturing the network-level spatial and temporal correlations among the various measurements.

Chapter 7

Validation

Constantinos Antoniou, Jordi Casas,
Haris Koutsopoulos, and Ronghui Liu

CONTENTS

This chapter provides a more detailed discussion of the validation principles outlined in Chapter 4. Section 7.1 introduces validation of traffic simulation models and puts the presented work in the context of the available literature, at the same time relating it to the key references on which their structures are based. Section 7.2 extends Table 4.2, focusing on how each of the presented measures of goodness-of-fit have been used by various researchers. Section 7.3 discusses statistical validation, focusing on hypothesis testing and tests of underlying structures. Application examples are presented in Section 7.4, while the discussion in Section 7.5 concludes the chapter.

7.1 INTRODUCTION

Model validation, i.e., the process of checking the extent to which a model replicates reality, is an essential step in the development and application of every traffic model. In this chapter, the role of validation within the scope of model development and application is presented and the framework for performing validation is discussed. This chapter draws to a large extent upon Toledo and Koutsopoulos (2004), Barcelo and Casas (2004), and Hollander and Liu (2008a), with additional material from several additional references.

Despite the increasing popularity of traffic simulation models, little attention has been given in the literature to model validation. The two types of validation approaches are visual and statistical (Rao and Owen, 2000). In visual validation, graphical representations of the outputs from real and simulated systems are displayed side by side to determine whether they can be differentiated. The visualization may be based on the animation modules available in most traffic simulation models.

Alternatively, plots of outputs (e.g., flows and speeds) may be generated. Turing tests (Turing, 1950; Schruben, 1980) may also be used. These tests involve presenting experts with sets of observed and simulated outputs. The result depends on whether the experts can correctly identify the individual sets. In any case, the process remains inherently subjective and heuristic. Statistical validation applies goodness-of-fit measures, confidence intervals, and statistical tests to quantify the similarities between real and simulated systems.

Many published validation studies are based on visual comparisons of outputs from real and simulated systems or on comparisons of simple descriptive statistics. For example, Abdulhai et al. (1999) plotted the observed and simulated headway distributions, lane usage breakdown, and flow density curves and calculated the mean relative percent errors of total demands, link flows, and lane usage breakdown. Other examples include Rakha et al. (1996), Bloomberg and Dale (2000), Hall et al. (2000), Rilett et al. (2000), Fellendorf and Vortisch (2001), and Lee et al. (2001).

In many cases, validation is limited to an isolated road section or traffic corridor, thus avoiding the more complex behaviors and interactions associated with network applications. Jayakrishnan et al. (2001) note that network-wide validation is necessary to ensure that interactions among models within a simulation framework are captured correctly.

Validation is often oriented at a specific model (often the acceleration model). For example, Benekohal (1991) focuses on the car-following model and Fellendorf and Vortisch (2001) separately and independently validate car-following and lane changing models. Rao et al. (1998) proposed a multilevel approach consisting of conceptual and operational validation. Conceptual validation focuses on the consistency of the simulation results with the theoretical foundations. For operational validation, the authors propose a two-level evaluation of the simulated data against real-world observations using statistical comparisons of the respective means and the distributions. They demonstrate the application of the proposed methods using platoon data collected from video recordings.

According to Law and Kelton (2000), the key methodological steps for building valid and credible simulation models are:

- Verification: determining that a simulation computer program performs as intended and is concerned with building the model correctly.
- Validation: determining whether a conceptual simulation model (as opposed to a computer program) is an accurate representation of the system under study; building the correct model.
- A model is credible when its results are accepted by the user and are used as an aid in making decisions. Animation is an effective way for an analyst to establish credibility.

Balci (1998) defines a successful simulation study as "one that produces a sufficiently credible solution that is accepted and used by decision makers." This implies the assessment of the quality of a simulation model through verification and validation. Verification usually implies running a simulation model under a variety of settings of input parameters and checking to see that the output is reasonable.

In some cases, certain measures of performance may be computed exactly and used for comparison. Animation can also be of great help for this purpose. For some types of simulation models (traffic models are good examples), it may be helpful to observe an animation of the simulation output to establish whether a computer model works as expected. In validating a simulation model, an analyst should keep in mind that:

- A simulation model of a complex system can only approximate the actual system. There is no such thing as an absolutely valid model of a system.
- A simulation model should always be developed for particular purposes and should be validated relative to measures of performance that will actually be representative of the purposes.
- Model development and validation should be performed hand-in-hand throughout an entire simulation study.

7.2 SELECTION OF GOODNESS-OF-FIT MEASURES

A number of goodness-of-fit measures can be used to evaluate the overall performances of simulation models (Table 7.1). Popular among them are the root mean square error (RMSE), root mean square percent error (RMSPE), mean error (ME), and mean percent error (MPE) statistics that quantify the overall errors of a simulator.

Percent error measures directly provide information on the magnitude of the errors relative to the average measurement. $RMSE$ and $RMSPE$ penalize large errors at a higher rate relative to small errors. ME and MPE indicate systematic under- or overprediction in the simulated measurements. These two statistics are most useful when applied separately to measurements at

Table 7.1 Measures of goodness of fit

Name	User
Percent error *(PE)*	Shaaban and Radwan (2005)
	MOEs: maximum queue length and travel distance in right-most lane
	Type of network: arterial with signalized intersections
	Type of data: cameras for collecting counts and turning percentages, maximum queue length, and travel distance for right-most lane
	Park and Qi (2005)
	MOE: average travel time on selected movements
	Type of network: actuated signalized intersection
	Type of data: traffic counts
	Merritt (2004)
	MOEs: queue and delay
	Type of network: arterial section
	Type of data: observed field data from traffic camera
Squared error *(SE)*	Ben-Akiva et al. (2004)
	MOE: link flows
	Type of network: freeway and arterials
	Type of data: real loop detector data
	Chu et al. (2004)
	MOEs: flow and travel time
	Type of network: freeway and arterials
	Type of data: real detector: 5-minute slice for freeways and 15-minute slice for arterials; probe vehicles for travel time
Mean error *(ME)*	Toledo and Koutsopoulos (2004)
	MOEs: speeds and metamodel parameters
	Type of network: freeway
	Type of data: measured counts and speeds from point sensors
Mean normalized error *(MNE)* or mean percent error *(MPE)*	Toledo et al. (2003)
	MOEs: link flows, point-to-point travel times, and queue lengths
	Type of network: mixed urban–freeway
	Type of data: measured link flows, point-to-point travel times, and queue lengths

Table 7.1 Measures of goodness of fit (Continued)

Name	User
	Toledo and Koutsopoulos (2004)
	MOEs: speeds and metamodel parameters Type of network: freeway Type of data: measured counts and speeds from point sensors
	Chu et al. (2004)
	MOEs: flow and travel time Type of network: freeway and arterials Type of data: real detector: 5-minute slice for freeways and 15-minute slice for arterials; probe vehicles for travel time
Mean absolute error (*MAE*)	Ma and Abdulhai (2002)
	MOE: link flows Type of network: freeway and arterials Type of data: turning counts at key signalized intersections
Mean absolute normalized error (*MANE*) or mean absolute error ratio (*MAER*)	Ma and Abdulhai (2002)
	MOEs: link flows Type of network: freeway and arterials Type of data: turning counts at key signalized intersections
	Kim and Rilett (2003)
	MOE: link flows Type of network: interstate Type of data: volume from pneumatic tubes and AVI OD
	Merritt (2004)
	MOEs: queue and delay Type of network: arterial section Type of data: field data observed from traffic camera
	Kim et al. (2005)
	MOE: travel time Type of network: signalized arterial operating under coordinated control system Type of data: travel times collected during morning peak (7:30 to 8:30 a.m.)

(continued)

Table 7.1 Measures of goodness of fit (Continued)

Name	User
Exponential mean absolute normalized error (*EMANE*)	Kim and Rilett (2004) MOE: link flows Type of network: interstate Type of data: volume data from pneumatic tubes and space mean speed from AVI sensors
Root mean squared error (*RMSE*)	Koutsopoulos and Habbal (1994) MOE: flows Type of network: main routes and arterials Type of data: HCM computations and simulated flows (SATURN) Toledo and Koutsopoulos (2004) MOEs: speeds and metamodel parameters Type of network: freeway Type of data: measured counts and speeds from point sensors Dowling et al. (2004) MOEs: capacity, speed, and travel time Type of network: freeway and arterials Type of data: synthetic: capacity calculated with HCM; real: speeds and travel times from detectors
Route mean squared normalized error (*RMSNE*) or route mean square percent error (*RMSPE*)	Hourdakis et al. (2003) MOEs: detector volumes, detector occupancies, and speeds Type of network: freeway Type of data: real detector data (5-minute slice) Toledo et al. (2003) MOEs: link flows, point-to-point travel times, and queue lengths Type of network: mixed urban–freeway Type of data: measured link flows, point-to-point travel times, and queue lengths Toledo and Koutsopoulos (2004) MOEs: speeds and metamodel parameters Type of network: freeway Type of data: measured counts and speeds from point sensors

Table 7.1 Measures of goodness of fit (Continued)

Name	User
	Ma and Abdulhai (2002)
	MOE: link flows Type of network: freeway and arterials Type of data: turning counts at key signalized intersections
GEH statistic	Barcelo and Casas (2004)
	MOE: flow Type of network: freeway Type of data: real detector (1-hour slice)
	Chu et al. (2004)
	MOEs: flow and travel time Type of network: freeway and arterials Type of data: real: 5-minute slice for freeways and 15-min- ute slice for arterials; probe vehicles for travel times
	Oketch and Carrick (2005)
	MOEs: flow, queue length, and travel time Type of network: urban Type of data: real detector (level of aggregation not available)
Correlation coefficient (r)	Hourdakis et al. (2003)
	MOEs: flows, occupancies, and speed Type of network: freeway Type of data: real detector (5-minute slice)
Theil's bias proportion (U^M)	Koutsopoulos and Habbal (1994)
	MOE: intersection delays Type of network: local including intersections Type of data: simulated flows (SATURN)
	Hourdakis et al. (2003)
	MOEs: flows, occupancies, and speed Type of network: freeway Type of data: real detector (5-minute slice)
	Toledo and Koutsopoulos (2004)
	MOEs: speeds and metamodel parameters Type of network: freeway Type of data: measured counts and speeds from point sensors

(continued)

Table 7.1 Measures of goodness of fit (Continued)

Name	User
	Barcelo and Casas (2004)
	MOEs: flows and occupancies Type of network: freeway Type of data: real detector (5-minute slice)
	Brockfeld et al. (2005)
	MOEs: flow, speed, and occupancies Type of network: freeway Type of data: real detector (5-second slice)
Theil's variance proportion (U^s)	Koutsopoulos and Habbal (1994)
	MOE: intersection delays Type of network: local including intersections Type of data: simulated flows (SATURN)
	Hourdakis et al. (2003)
	MOEs: flows, occupancies, and speed Type of network: freeway Type of data: real detector (5-minute slice)
	Toledo and Koutsopoulos (2004)
	MOEs: speeds and metamodel parameters Type of network: freeway Type of data: measured counts and speeds from point sensors
	Barcelo and Casas (2004)
	MOEs: flow and occupancies Type of network: freeway Type of data: real detector (5-minute slice)
	Brockfeld et al. (2005)
	MOEs: flow, speed, and occupancies Type of network: freeway Type of data: real detector (5-second slice)
Theil's covariance proportion (U^c)	Koutsopoulos and Habbal (1994)
	MOE: intersection delays Type of network: local including intersections Type of data: simulated flows (SATURN)

Table 7.1 Measures of goodness of fit (Continued)

Name	User
	Hourdakis et al. (2003)
	MOEs: flows, occupancies, and speed Type of network: freeway Type of data: real detector (5-minute slice)
	Toledo and Koutsopoulos (2004)
	MOEs: speeds and metamodel parameters Type of network: freeway Type of data: measured counts and speeds from point sensors
	Barcelo and Casas (2004)
	MOEs: flow and occupancies Type of network: freeway Type of data: real detector (5-minute slice)
Theil's inequality coefficient (U)	Koutsopoulos and Habbal (1994)
	MOE: intersection delays Type of network: local including intersections Type of data: simulated flows (SATURN)
	Ma and Abdulhai (2002)
	MOE: link flows Type of network: freeway and arterials Type of data: turning counts at key signalized intersections
	Hourdakis et al. (2003)
	MOEs: flows, occupancies, and speed Type of network: freeway Type of data: real detector data (5-minute slice)
	Toledo and Koutsopoulos (2004)
	MOEs: speeds and metamodel parameters Type of network: freeway Type of data: measured counts and speeds from point sensors
	Barcelo and Casas (2004)
	MOEs: flow and occupancies Type of network: freeway Type of data: real detector (5-minute slice)

(continued)

Table 7.1 Measures of goodness of fit (Continued)

Name	User
	Brockfeld et al. (2005):
	MOEs: flow, speed, and occupancies Type of network: freeway Type of data: real detector (5-second slice)
Kolmogorov-Smirnov test	Kim et al. (2005)
	MOE: travel time Type of network: signalized arterial, operating under coordinated control system Type of data: travel times collected during morning peak (7:30 to 8:30 a.m.)
Moses and Wilcoxon tests	Kim et al. (2005)
	MOE: travel time Type of network: signalized arterial, operating under coordinated control system Type of data: travel times collected during morning peak (7:30 to 8:30 a.m.)

MOE = margin of error.

x_i = simulated measurement.

y_i = observed measurement.

N = number of measurements.

$\overline{x}, \overline{y}$ = sample average.

σ_x, σ_y = sample standard deviation.

Source: Adapted from Hollander, Y. and Liu, R. 2008a. With permission.

each time and space point rather than to all measurements jointly. This way, they provide insight into the spatial and temporal distribution of errors and help identify deficiencies in a model.

Another measure that provides information on relative error is Theil's inequality coefficient U (Theil, 1961). $U = 0$ implies perfect fit between the observed and simulated measurements. $U = 1$ implies the worst possible fit. Theil's inequality coefficient may be decomposed to three proportions of inequality: the bias (U^M), variance (U^S), and covariance (U^C) proportions. The bias proportion reflects the systematic error. The variance proportion indicates how well the simulation model replicates the variability in observed data. These two proportions should be as small as possible. The covariance proportion measures the remaining error and therefore should be close to one. Note that if the different measurements are taken from nonstationary processes, the proportions can be viewed only as indicators of the sources of error.

7.3 STATISTICAL VALIDATION

7.3.1 Hypothesis testing and confidence intervals

Classic hypothesis tests (two-sample t, Mann-Whitney, and two-sample Kolmogorov-Smirnov) and confidence intervals may also be used. Law and Kelton (2000) suggest the use of confidence intervals that provide richer information compared to statistical tests for the validation of complex simulation systems.

Two-sample tests assume that both sets of outputs are independent draws from identical distributions (iids). Therefore, these tests should be performed separately for each time–space measurement point. If the number of observations at each time–space point is not sufficient to yield significant results, observations from appropriate time intervals (such that the iid assumption holds, at least approximately) may be grouped together.

The standard two-sample t-test also assumes that the two distributions (observed and simulated) are normal and share a common variance. These assumptions, in particular the variance equality, may be unrealistic in the context of traffic simulation. Law and Kelton (2000) propose an approximate t solution procedure based on Scheffé (1970) that relaxes the variance equality assumption. To test for the equality of the mean of observed and simulated measurements, $H_0 : Y^{sim} = Y^{obs}$ against $H_1 : Y^{sim} \neq Y^{obs}$ at the α significance level, reject H_0 if:

$$\frac{|Y^{sim} - Y^{obs}|}{\sqrt{\frac{(s^{sim})^2}{n^{sim}} + \frac{(s^{obs})^2}{n^{obs}}}} \geq t_{\alpha/2, \hat{f}} \tag{7.1}$$

Y^{obs}, Y^{sim}, s^{obs}, and s^{sim} are the sample means and standard deviations of the observed and simulated measurements, respectively. n^{obs} and n^{sim} are the corresponding sample sizes. $\hat{f}$ is the modified number of degrees of freedom given by:

$$\hat{f} = \frac{\left(\frac{s^{sim}}{n^{sim}} + \frac{s^{obs}}{n^{obs}}\right)^2}{\frac{(s^{sim})^4}{(n^{sim})^2(n^{sim}-1)} + \frac{(s^{obs})^4}{(n^{obs})^2(n^{obs}-1)}} \tag{7.2}$$

The corresponding $(1 - \alpha)$ confidence interval is given by:

$$Y^{sim} - Y^{obs} \pm t_{\alpha/2, \hat{f}} \sqrt{\frac{(s^{sim})^2}{n^{sim}} + \frac{(s^{obs})^2}{n^{obs}}} \tag{7.3}$$

The above approaches can be used for the analysis of individual time–space points. However, the behaviors of traffic networks in many applications are autocorrelated and nonstationary, because of time-varying travel demands and traffic dynamics (congestion build-up and dissipation). Therefore, measurements at different space–time points cannot be considered independent draws from identical distributions and the above methods cannot be used to test the overall validity of the simulation.

Joint hypothesis tests that better reflect the dynamics of the system are more appropriate. Kleijnen (1995) recommends the application of Bonferroni's inequality to test multiple hypotheses jointly at a prescribed significance level α. Let α_n be the significance level at each individual time–space point n. An upper bound for the simultaneous significance level α at the network level is given by:

$$\alpha \leq \sum_{n=1}^{N} \alpha_n \tag{7.4}$$

N is the number of measurement points (over time and space). Equation (7.4) holds under very general conditions. In practice, the significance levels at each time–space point n are usually set to $\alpha_n = \alpha/N$. α_n is then used as the level of significance independently at each time–space location to perform hypothesis testing or develop confidence intervals. Bonferroni's inequality can be used similarly to create composite tests and joint confidence intervals for multiple single-valued MOPs.

In practice, this technique may be applied only to a small number of measurement points N. For large N, the corresponding significance levels α_n become small and the confidence intervals increase to the point where it is difficult to reject any model as invalid.

Another limitation of Bonferroni's inequality is that it is too conservative, especially for highly correlated test statistics, hence resulting in a high probability of a Type II error, i.e., failure to reject a false null hypothesis. Holm's test (1979) is a more powerful version of Bonferroni's test and works better when the tests are correlated and several null hypotheses are false. Other methods that improve the power of the test have also been proposed (Shaffer, 1995).

An alternative sequential test adapted from Rao et al. (1998) involves a two-sample Kolmogorov-Smirnov (K-S) test for each measurement (i.e., time–space point or MoP) and recording the corresponding p values. A one-sided t-test is then conducted to see whether the mean of these p values is smaller than the desired significance level (i.e., the model is invalid).

As discussed above, two-sample tests such as the K-S can be used to determine whether simulated and observed measurements are drawn from the same distribution and thus examine the validity of a simulation.

The application of the K-S test is straightforward for the case of a single MoP. Rao et al. (1998) suggest the use of a two-dimensional two-sample K-S test to validate a pair of MoPs jointly. This test is useful because MoPs (e.g., flows and headways) are usually correlated. Although the test involves several steps, it is simple to implement and apply. Details can be found in Fasano and Franceschini (1987) and Press et al. (1992), who also provide who an approximate function for the test p values.

In the above discussion, no assumption was made about the nature of the input data. If the input data are known (trace-driven simulation), Kleijnen (1995 and 1999) proposes a regression procedure to validate the simulation model using an F-test of the joint equality of the means and variances of the real and simulated measurements. Let us assume that there are N different input data sets (common to the simulated and true systems). For pairs of observations (y_n^{obs}, y_n^{sim}), $n = 1, \ldots, N$ the following regression is performed:

$$\left(y_n^{obs} - y_n^{sim} \right) = \beta_0 + \beta_1 \left(y_n^{obs} - y_n^{sim} \right) + \varepsilon_n \tag{7.5}$$

The hypothesis that the observed and simulated outputs are drawn from identical distributions is tested with the null $H_0 : \beta_0 = 0$ and $\beta_1 = 0$.

7.3.2 Tests of underlying structure

Toledo and Koutsopoulos (2004) propose another approach, particularly suitable for validating traffic simulation models with limited data. The method is based on developing metamodels that capture the underlying relations among important traffic variables of interest (e.g., speed–flow relationships, time evolutions of flows) and statistically testing the hypothesis that the metamodel parameters are equal in the simulated and observed data.

The validation proceeds by using the outputs from the real and simulated systems to estimate two separate metamodels that describe the structures of these outputs. The choice of the appropriate metamodel depends on the nature of the application and relationships among variables established by the traffic flow theory. Statistical tests for the equality of coefficients across the two metamodels are used to validate the simulation model. The equality of the models is then tested with the null hypothesis $H_0 : \beta^{obs} = \beta^{sim}$ against $H_1 : \beta^{obs} \neq \beta^{sim}$ using an F-test.

The test uses two models: restricted and unrestricted. The restricted model that forces the equality of parameters of the two metamodels is estimated with the combined data set (real and simulated observations). The unrestricted model is the combination of two separate models, one estimated with real data and the other with simulated data. The test statistic is calculated:

$$F_{K,N^{sim}+N^{obs}-2K} = \frac{\frac{ESS^R - ESS^{UR}}{K}}{\frac{ESS^{UR}}{N^{sim}+N^{obs}-2K}} \tag{7.6}$$

ESS^R and ESS^{UR} are the sums of squared residuals of the restricted and the unrestricted models, respectively. They are calculated as $ESS^R = ESS^{com}$ and $ESS^{UR} = ESS^{obs} + ESS^{sim}$. ESS^{com}, ESS^{obs}, and ESS^{sim} are the statistics calculated from the models estimated with the combined, real, and simulated data, respectively. N^{obs} and N^{sim} are the numbers of observations in the real and simulated data, respectively. K is the number of parameters in the model. This method provides great flexibility in validating simulation models and in many cases overcomes the difficulty of limited data.

In many traffic studies, measurements may only be available for only a few days. This poses a problem for most of the methods discussed above because limited data curbs the development of statistically significant test statistics and goodness-of-fit measures. Moreover, disaggregate data (e.g., observations of individual vehicles, one-minute sensor readings) can be used when fitting the metamodel without having to aggregate observations (e.g., to 5-minute intervals) as would be required if real and simulated outputs were directly compared.

Another advantage is the flexibility in choosing the functional form of the metamodel to accommodate different modeling needs. For example, a metamodel describing an MoP as a function of time (e.g., time-dependent speed) can be used to test whether the formation and dissipation of congestion are captured correctly. Another metamodel can be formulated to test the realism of the underlying fundamental diagram.

7.4 APPLICATION EXAMPLES

This section provides some application examples demonstrating the above concepts at different levels.

7.4.1 Strategic level

Toledo and Koutsopoulos (2004) demonstrated the application of the methods discussed above in the validation of a microscopic simulation model using 3-hour morning peak sensor data from a section of the M27 freeway in Southampton, England, shown in Figure 7.1. The observed data include counts and speeds for 5 days at 1-minute intervals at the locations indicated in the figure. The simulation outputs include similar measurements from 10 runs. For the purpose of this demonstration, we consider the measurements at sensor 3.

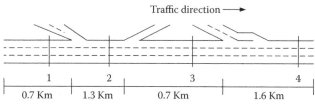

Note: Figure not drawn to scale

Figure 7.1 M27 freeway section in Southampton, England. (*Source:* Toledo, T. and Koutsopoulos, H.N. 2004. *Transportation Research Record,* 1876, 142–150. With permission.)

Toledo and Koutsopoulos (2004) first calculated goodness-of-fit statistics for the sensor speeds based on measurements at 1-minute intervals. The results are summarized in Table 7.2 and indicate a reasonable overall fit, with a small bias and good ability to replicate the variability in the data. One limitation is that these statistics do not evaluate the ability of a simulation model to represent the dynamics of traffic behavior.

Next, Toledo and Koutsopoulos (2004) performed two-sample tests focused on a specific time interval. Measurements were grouped in 15-minute intervals to compile enough observations for the statistical tests. The assumption was that within each interval observations were independent draws from the same distribution. The tests were performed separately for each interval. For example, Table 7.3 summarizes the test results for the third time period.

The null hypothesis $H_0 : Y^{sim} = Y^{obs}$ is rejected at 95% confidence. The corresponding confidence interval is $-4.72 \leq Y^{sim} - Y^{obs} \leq -1.86$. Note that while the goodness-of-fit statistics show good overall fit the focused analysis of specific time intervals reveals weaknesses of the simulation model in capturing traffic dynamics. This result illustrates the need to use statistical methods that are as detailed as possible.

Table 7.2 Goodness-of-fit statistics for Southampton network

Statistic	Value
RMSPE (%)	5.08
RMSE (Km/h)	5.09
MPE (%)	−0.66
ME (Km/h)	−0.83
U (Theil's inequality coefficient)	0.024
U^M (bias proportion)	0.135
U^S (variance proportion)	0.023
U^C (covariance proportion)	0.842

Table 7.3 Two-sample *t*-test for third time interval

	Observed data	Simulated data
Mean	108.77	105.49
Variance	11.53	56.09
Observations	75	150
t statistic	4.52	
t critical value	1.97 (95% confidence)	

Individual tests over all time periods may yield conflicting results and application of Bonferroni's inequality may be too conservative. Therefore, Toledo and Koutsopoulos (2004) tested the validity of the model with the application of hypothesis testing on metamodels. They first developed metamodels that described the fundamental diagrams underlying the two sets of data. The functional form of the Pipes-Munjal model (Pipes, 1967) was selected:

$$V(t) = V_f \left[1 - \left(\frac{\rho(t)}{\rho_{jam}} \right)^{\beta} \right] + \varepsilon(t) \tag{7.7}$$

$V(t)$ and $\rho(t)$ are the traffic speed and density at time t, respectively. V_f denotes the free flow speed, ρ_{jam} is the jam density, and β represents the underlying parameters of the model. $\varepsilon(t)$ is an error term. The observed data and the simulated data used for estimation and the corresponding estimated metamodel regression lines are shown in Figure 7.2. For the purpose of statistical testing, a third metamodel was estimated using the combined data set, including both observed and simulated data.

The *F*-test described above may be used to test the validity of the simulation output:

$$F_{3,2694} = \frac{52034.3 - 47698.4/3}{47698.4/1800 + 900 - 6} = 81.63 \tag{7.8}$$

The critical value of the *F* distribution with 3.2694 degrees of freedom at the 95% confidence level is 8.53, and the hypothesis that the parameters of the observed and simulated metamodels are equal can be rejected. Further insight into the performance of a simulation model can be derived by developing a piecewise linear metamodel describing temporal changes in traffic speeds:

$$V(t) = V_0 + \alpha_1 t + \alpha_2 (t - \beta_1) x_1(t) + \cdots + \alpha_i (t - \beta_{i-1}) x_{i-1}(t)$$
$$+ \cdots + \alpha_N (t - \beta_{N-1}) x_{N-1}(t) + \varepsilon(t) \tag{7.9}$$

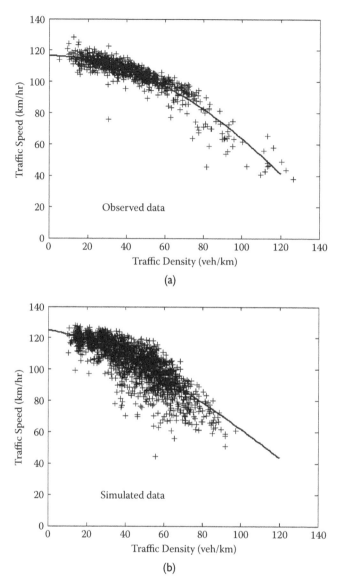

Figure 7.2 Observed and simulated speed density data and fitted metamodels. (*Source:* Toledo, T. and Koutsopoulos, H.N. 2004. *Transportation Research Record,* 1876, 142–150. With permission.)

$V(t)$ is the traffic speed at time t. V_0, α, and β are parameters of the model. α_i captures the change in the slope of section i from the previous section. β_i is the boundary point between sections i and $i + 1$. $\varepsilon(t)$ is an error term. $x_i(t)$ is defined as:

$$x_i(t) = \begin{cases} 1 & \text{if } t \geq \beta_i \\ 0 & \text{otherwise} \end{cases} \quad i = 1,\ldots,N-1 \tag{7.10}$$

Metamodels of this form were estimated for the real data, simulated data, and combined data set using the Gallant-Goebel estimator for non linear least squares problems with time series data (Galland and Goebel, 1976). The number of linear segments was set at three by maximizing $\bar{R}^2$ in the observed data nonparametrically. The regression lines of the estimation results are shown in Figure 7.3.

For this representation of the data, an F-test for the equality of coefficients in the data yields the statistic $F_{6,2688} = 18.31$. The critical value of the F distribution at the 95% confidence level is 3.67, and so the null hypothesis can be rejected, i.e., with respect to the time–speed space, the simulation model is not a valid representation of the real system. Similar tests assuming partial equality of the parameters may also be conducted to better explain the sources of the differences between the two data sets. For example, a test for the equality of the boundaries of the linear pieces $(H_0 : \beta_1^{obs} = \beta_1^{sim}$ and $\beta_2^{obs} = \beta_2^{sim})$ will reveal whether the build-up and dissipation of congestion occur at the same times in the two data sets.

7.4.2 Tactical level

Casas et al. (2010) present a framework for validation of traffic models based on trajectory data. The primary results of the current working process focus on tactical trajectory data and as a future research project will be extended to strategic trajectory data. Trajectory data in this chapter can be seen as a sequence of physical points that represents a path from an origin point to a destination point. Let $p = (p_1, p_2, \ldots, p_n)$ be an n-dimensional vector, where $p_i = x_i, y_i$.

Figure 7.4a depicts all real trajectories from on-ramp lane to left-most lane, where the x-axis represents x_i and the y-axis y_i.

Observed and simulated trajectories can be seen as functional data and several statistical methods are available for their analysis. Functional data analysis (FDA) is a collection of statistical techniques that try to adapt the classical statistical methods for the analysis of functional data. The

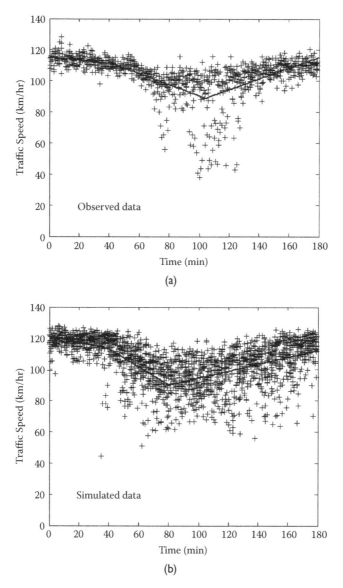

Figure 7.3 Observed and simulated time speed data and fitted metamodels. (*Source:* Toledo, T. and Koutsopoulos, H.N. 2004. *Transportation Research Record,* 1876, 142–150. With permission.)

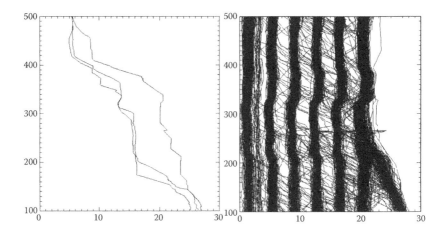

Figure 7.4 (a) Representation of trajectories from on-ramp lane to left-most lane. (b) Representation of all trajectories captured in real data set. (*Source:* Casas, J. et al. 2010. Proceedings of Traffic Flow Theory Summer Meeting and Conference, Annecy, France, July 7–9. With permission.)

measures of mean, standard deviation, and correlation can be applied to functional data samples. The main difference is the requirement to consider that these statistics are working in a different vector space or different norm: norm-L_2 (L_2 is the Hilbert norm). A book by Ramsay and Silverman (1997) offers a broad perspective of the available methods and case studies in functional data analysis.

In this work, a cluster analysis is proposed to "classify" the trajectories and determine whether the classification corresponds to a different type of trajectory, either in terms of origin and destination lane or traffic regimes or conditions (such as free flow, congestion, and the transition between free flow and congestion). Complementarily, this cluster analysis could be performed using a prior classification of the trajectories based on origin lane and destination lane. Figure 7.4b depicts the patterns determined by the origin and destination lanes whose prior classifications make sense to apply.

To deal with clustering techniques, a concept of distance between two trajectories must be specified. A standard definition of distance between two generic trajectories f and g as a function of parameter x (i.e., time, position, distance traveled, etc.) (Bruno and Greco, 2008; Nanni and Pedreschi, 2006) between the intervals $[x_{\min}, x_{\max}]$ is the integral:

$$d(f,g) = \int_{x_{\min}}^{x_{\max}} \|f(x) - g(x)\| dx \qquad (7.11)$$

where $\|\cdot\|$ denotes the norm. That is based on the L^2 norm:

$$\|x\| = \langle x, x \rangle = \int x^2(t)dt \qquad (7.12)$$

After a measure of distance has been defined, various clustering methods are evaluated. A large range of possibilities can be evaluated (Johnson and Wichern, 1998). We propose application of multidimensional scaling (MDS) that tries to find a configuration in $\mathbb{R}^q$ in such a way that distances between points try to reflect the distances between trajectories. More details on MDS theory and applications are available in Young and Hamer (1987).

The other method is the dendrogram—a tree that shows the clusters produced by a hierarchical clustering. Most methods are based on measures of similarity or dissimilarity between sets of observations, in this case the distance proposed above.

After the clustering of all trajectories has been calculated, the next step is the validation of simulated trajectories with the observed ones. To perform validation, two methods based on input data have been proposed.

When the aim of a study is to check whether one vehicle trajectory fits into any of the calculated clusters, a graphical approach is used to decide whether this trajectory fits into a cluster. The graphical approach plots the confidence interval of the clusters and then the user manually calculates whether the vehicle trajectory fits. When available data consist of a set of observed trajectories and a set of simulated trajectories, the proposed method is an ANOVA test for functional data proposed by Cuevas et al. (2003).

7.5 DISCUSSION

Calibration and validation are complementary processes that are key to the success of any model application. It is important to find and utilize good practices in a way that is accessible to researchers and practitioners to bridge the gap between state-of-the-art and state-of-the-practice. This chapter presents applications and a systematic review of validation efforts from the literature, focusing on the selection of goodness-of-fit measures and their applicability and state-of-the-art statistical validation concepts.

A key point is the large discrepancy that seems to exist between the state of the art and the state of the practice. Despite a large body of literature about validation, the technique in practice is often restricted to a very rudimentary effort or is not performed at all. If validation is not properly performed, overfitting or incorrect calibration may go unnoticed and will ultimately produce an unreliable model. The need for calibration is particularly relevant in the field of transport simulation, where complex, highly nonlinear models with many degrees of freedom are used and multiple combinations of calibrated inputs and parameters could produce similar calibration results and very different validation results.

While a large number of goodness-of-fit measures have been applied to validation, there is still no consensus on whether a single measure or a subset would be better suited to the task. Perhaps each measure sheds light on a different aspect of the problem, and therefore, the measures should be computed and presented in parallel. Similarly, it is not clear whether certain traffic measurements are more suited to validation or whether suitability depends on the type of simulator model or the application at hand. These and related questions must be answered by future research.

Chapter 8

Conclusions

Christine Buisson, Serge Hoogendoorn,
and Winnie Daamen

CONTENTS

Transportation network simulation tools are widely used. They permit end users to assess a priori or a posteriori numerous infrastructure designs and changes therein. Simulation is also used to determine the impacts of traffic management methods, driver support systems, and automated driving on, e.g., roadway capacities and travel times. In the past decade, the types of assessment criteria have expanded with the objectives to estimate congestion relief and quantify the environmental and traffic safety impacts. The range of applications of traffic simulation software has broadened significantly.

Notwithstanding the expanded application perspectives of traffic simulation, the validity of the simulation tools used is not undisputed. In particular, the so-called predictive validity of simulation models, that is, their ability to predict network flow operations in situations for which they have not been explicitly calibrated has been questioned. One reason is the complexity of the modeling task.

This complexity stems first and foremost from the fact that our study subject, i.e., a transportation system, does not always adhere to strict physical and mathematical laws. Surely, vehicles are conserved and their dynamic actions adhere to simple kinematic equations and laws of physics, but driver behavior, which is a key factor in every system, is difficult to predict. The inductive approach to theory building and modeling provides us with mathematical constructs that are at best useful for the simulation tasks at hand but are generally not applicable to all modeling tasks we can devise.

Let us take a closer look at the inherent complexity of modeling traffic in transportation networks. This complexity arises from a number of causes. As stated earlier, humans play a critical role; their behaviors and interactions are very complex to model mathematically due to factors such as heterogeneities among drivers, intra-driver changes, errors in decision making, and decision making under uncertainty.

Physical limitations of vehicles and infrastructures are important factors. A model should consider the physical limitations of a transportation system, such as lane capacity and maximum acceleration rate.

Due to the many internal factors (e.g., travel and driving behavior, occurrence of accidents) and external factors (e.g., weather conditions, manmade or natural hazards), determining the behavior of a system based on the uncertain impacts of these factors on operations is a very difficult task.

The role of the human factor (e.g., interaction with traffic management measures) in this interaction is complex and still poorly understood. In particular, the human responses in terms of compliance with instructions, confidence in information provided by a system, driver workload, attention levels, and distractions deserve a substantial amount of additional research. A many-faceted transportation system involves a large number of interacting components that often self-organize into stable or unstable unpredictable states and produce chaos-like conditions.

Some of the complexities of these transportation phenomena are only moderately reflected in simulation tools. The models are usually composed of various submodels and encompass only a part of the observed stochasticity in a transportation system and its constituent elements. Each individual submodel is often already complex and the interactions among submodels make the overall system even more complex.

As an example, consider the link between a longitudinal behavior model that explains the lengths of the gaps between cars travelling on a road and a lateral behavior model that may predict that a given gap will be accepted to change lanes. The resulting lane changing maneuver divides the length of this gap by a factor of about two, thus modifying the longitudinal behavior and reducing the probability that a lane changing maneuver may be further operated.

Moreover, most of the submodels will contain stochastic elements. As a result, general conclusions cannot be drawn on the basis of a single run because it is only a single realization of a random process. Conclusions must result from an analysis of the simulation results distributions because each run corresponds to a different random realization.

8.1 SUMMARY

The validity of any simulation model can be assessed at different levels. In the first stages of model development, one often looks for plausible outcomes by considering the face validity of the model. For practical applications, however, face validity is often not enough and comparison with field observations is required. This generally is achieved first by calibration of the model, followed by testing the model's predictive validity. Predictive validity includes necessarily a comparison of the simulation data with empirical data but not on the level of detail of validity.

In both validation processes, data availability and data quality are key requirements. For many transportation model problems, data are often noisy and incomplete, and their collection is often costly and cumbersome. The incompleteness of data may in some cases be partly overcome through the use of data processing techniques such as filtering and data fusion even in cases where the spatial and temporal aggregation periods are different or the data differ in nature. However, some data are still difficult to collect, even with the availability of new collection technologies. For example, (dynamic) origin destination matrices are, on the one hand, essential inputs for any simulation model, but they are also difficult to observe or infer from available data sources. Furthermore, some data are simply not measurable by nature. Examples are drivers' motivations and preferences that may be assessed only by observing driving behavior.

Much debate has surrounded the validity of transportation simulation models. One reason is that policy makers who base their decision making on the outcomes of the simulation models often do not fully understand the complex nature and limitations of the models. The seemingly high level of detail and realism that microscopic simulation models appear to have—just look at the fancy three-dimensional animations that these models can produce—provide end users and policy makers with the often false impression that models truly provide predictively valid outcomes at the microscopic level.

On the other hand, the end users of complex models often face large challenges in using them within the limitations of the projects for which they were designed, for example, budget and time constraints and limited data availability. The lack of useful guidelines and tools for model calibration and validation is another concern.

Specifically, in relation to the lack of useful guidelines and tools, this book paves the way for an increase in the confidence in simulation tools by providing methods for sensitivity analyses, calibration and validation, and tools for their appropriate application. To this end, it describes the various tasks of calibration and validation and explains how to best use the available data. It also presents the sensitivity analysis method that permits restriction of the number of parameters to optimize during the calibration phase.

With respect to the data needed for simulation model application (preparation of model input and compiling data needed for calibration), this book discusses the typical issues of data error in transportation system data and how these errors can impact simulation results. The book also presents ways to complete missing data and correct existing data, and also provides methods to unify data from various data sources with similar and different semantics. This three-step process achieves a better quality of data for model input, calibration, and validation.

The first step consists of sensitivity analyses for determining which parameters are the most important to adjust. This method is new in the transportation field and an exhaustive presentation of this methodology is given in the book. Calibration is the second step. It identifies parameter values that minimize the differences between simulation results and observations.

Model calibration and validation rely on common and unambiguous definitions of the differences between observations and simulation results. In this book, this comparison involves a two-step decision process. The first choice is purpose-related and concerns the careful definition of the relevant measures of performance (MoPs). This choice is, of course, directly related to the simulation task at hand and is thus problem-specific.

Since transportation system improvements in many cases are intended to act on particular aspects of system characteristics (e.g., pollution, congestion levels, reliability, safety), the MoPs must be defined to allow the end user to quantify the impacts of the enhancement of a particular system characteristic and agree with the objectives of the scenarios. In other words, the considered simulation tool should be able to reproduce the MoPs with a sufficient level of accuracy. The various MoPs are described in detail and linked with measurement methods.

The second choice is the definition of the goodness-of-fit (GoF) indicators. These metrics compare the observed and simulated MoP values. The choice of this metric has an impact on the results. This book presents various GoF methods found in the literature and analyzes their advantages and pitfalls.

For calibration, the end user must decide which technique is best for determining the model parameters. In practice, this is often done by trial and error. In this book, we present methods to determine the parameter values so that some error measures are minimized. These intelligent optimization procedures can be used to replace trial-and-error or brute-force methods that require too much computation time. A comprehensive list of all applicable procedures is provided so that the reader may choose among the various available approaches.

8.2 IMPLICATIONS

While this book details various methodologies for data collection, sensitivity analysis, calibration, and validation, the question is what benefits will result from the application of these methods. First, the use of better data improves input data quality. By performing calibration and validation steps, the quality of simulation results will most likely be higher and allow more accurate predictions of future situations. This way, better designs and more efficient traffic management measures will result. Scientific implications relate to the availability of better data, leading to more extensive calibration of models and corresponding improvements of the models.

8.3 FUTURE RESEARCH

Future research directions and operational developments are needed to further the state-of-the-art of simulation presented in this book. Research efforts must be directed to finding methods to cope with multiple, incomplete, and erroneous data. This effort is needed to ensure the efficient use of increasing mobile sources of data. Because mobile data collection devices are linked to travelers and not to vehicles, they provide inherently multimodal information.

Easier-to-use and faster calibration methods can result from two lines of research: (1) better understanding of sensitivity analysis methods to identify the most important parameters and (2) transposition of parameter optimization procedures existing in other simulation areas.

One of the essential characteristics of simulation models is their stochasticity. Stochasticity also occurs in reality. However, a full recognition of its importance and how to deal with it remain to be achieved. For example, a methodology to compare various replications of simulation runs with identical parameters and entry data is still under construction, a research area that can also gain from transposition of other simulation domains results. The definition of a "representative day" is still under discussion. Every day differs from the one that precedes it and the one that follows it. For that reason, the "representative day" is still undetermined, which raises the question to what data the simulation results should be prepared.

Other research directions concern the improvement of the use of simulation tools. The MULTITUDE project has produced guidelines for end users of simulation models. Concerns about the predictive validity of our models directly relating to the range of applications for which the current simulation models are in fact suitable are key drivers to further investigate traffic and transportation processes. Based on the inductive nature of this research domain, gaining more insight into the underlying traffic operational and physical processes will first and foremost be aided by developing new data collection techniques and also through improved data processing, calibration and validation methods, and new mathematical modeling techniques. The following directions should be investigated more thoroughly:

- Enhanced understanding of the roles of driver heterogeneity, behavior adaptation effects, and other factors in mixed (fast and slow) traffic situations.
- Theory and modeling of interactions of driving behavior and in-car technology, cooperative driving, and other measures.

With respect to the use of simulation tools, better understanding—both by researchers and by the end users—of the modeling paradigms on which the simulation models are based is a prerequisite. This understanding would reduce the risk of choosing inappropriate models, which is often observed in practice. Although it may sound elementary, not every model is applicable to every task and end users must understand models' limitations. In fact, in many cases, the use of a single model may not be appropriate and an end user may achieve better results from a combination of models. In addition, the increasing complexity of commercial models implies that end users will not always apply the models correctly.

In many countries, guidelines are available to support performing simulation studies. It would be of great use to cross-compare these guidelines and analyze the benefits and drawbacks of the different approaches. Furthermore, professionals can be better supported by providing data sets and tools for model calibration and validation.

References

Abdulhai B., Sheu, J.B., and Recker, W. 1999. Simulation of ITS on the Irvine FOT Area Using PARAMICS 1.5 Scalable Microscopic Traffic Simulator Phase I. Model Calibration and Validation. California PATH Research Report UCB-ITS-PRR-99-12.

Abdullah, S., Khalid, M., Yusof, R. et al. 2007. Comparison of feature extractors in license plate recognition. In *First Asia International Conference on Modeling and Simulation*, pp. 502–506.

Ahmed, K.I. 1999. Modeling drivers' acceleration and lane changing behavior. PhD thesis. Cambridge, MA: Massachusetts Institute of Technology.

Al-Battaineh, O. and Kaysi, I.A. 2007. Genetically optimized origin–destination estimation (GOODE) model: application to regional commodity movements in Ontario. *Canadian Journal of Civil Engineering*, 34(2), 228–238.

Alexopoulos, C. and Seila, A. 1998. Output data analysis. In J. Banks, Ed., *Handbook of Simulation: Principles, Methodology, Advances, Applications, and Practice*. New York: Wiley.

Alicandri, E. 1994. The Highway Driving Simulator (HYSIM): the next best thing to being on the road. *Public Roads*, 57(3), 19–23.

Anagnostopoulos, C., Anagnostopoulos, I., Loumos, V. et al. 2006. A license plate recognition algorithm for intelligent transportation system applications. *IEEE Transactions on Intelligent Transportation Systems*, 7(3), 377–392.

Antoniou, C., Balakrishna, R., Koutsopoulos, H.N. et al. 2009. Off-line and on-line calibration of dynamic traffic assignment systems. Presented at 12th IFAC Symposium on Control in Transportation Systems.

Antoniou, C., Ben-Akiva, M., and Koutsopoulos, H.N. 2007. Nonlinear Kalman filtering algorithms for on-line calibration of dynamic traffic assignment models. *IEEE Transactions on Intelligent Transportation Systems*, 8(4), 661–670.

Ashok, K. 1996. Estimation and prediction of time-dependent origin–destination flows. PhD thesis. Cambridge: Massachusetts Institute of Technology.

Ashok, K. and Ben-Akiva, M. 1993. Dynamic origin–destination matrix estimation and prediction for real-time traffic management systems. *In Proceedings of 12th International Symposium on Transportation and Traffic Theory*, pp. 465–484.

Ashok, K. and Ben-Akiva, M. 2000. Alternative approaches for real-time estimation and prediction of time-dependent origin-destination flows. *Transportation Science*, 34, 21–36.

ATAC. 2009. Non-Intrusive Traffic Detection Comparison Study. Final Report for North Dakota Department of Transportation prepared by Advanced Traffic Analysis Center at North Dakota State University.

Auberlet, J., Pacaux, M., Anceaux, F. et al. 2010. The impact of perceptual treatments on lateral control: a study using fixed-base and motion-base driving simulators. *Accident Analysis and Prevention*, 42(1), 166–173.

Balakrishna, R. 2002. Calibration of the demand simulator within a dynamic traffic assignment system. Master's thesis. Cambridge: Massachusetts Institute of Technology.

Balakrishna, R. 2006. Off-line calibration of dynamic traffic assignment models. PhD thesis. Cambridge: Massachusetts Institute of Technology.

Balakrishna, R., Antoniou, C., Ben-Akiva, M. et al. 2007. Calibration of microscopic traffic simulation models: methods and application. *Transportation Research Record*, 1999, 198–207.

Balakrishna, R., Ben-Akiva, M., and Koutsopoulos, H.N. 2007. Off-line calibration of dynamic traffic assignment: simultaneous demand–supply estimation. *Transportation Research Record*, 2003, 50–58.

Balci, O. 1998. Verification, validation, and testing. In J. Banks, Ed., *Handbook of Simulation*. New York: Wiley, pp. 335–393.

Barceló, J. and Casas, J. 1999. The use of neural networks for short-term prediction of traffic demand. In *Proceedings of 14th International Symposium on Transportation and Traffic Theory*. New York: Pergamon Press.

Barceló, J. and Casas, J. 2002. Dynamic network simulation with AIMSUN. Proceedings of International Symposium on Transport Simulation, Yokohama.

Barceló, J. and Casas, J. 2004. Methodological notes on the calibration and validation of microscopic traffic simulation model. Proceedings of 83rd Annual Meeting. Washington, D.C.: Transportation Research Board.

Barceló, J., Montero, L., Bullejos, M. et al. 2012. A Kalman filter approach for the estimation of time-dependent OD matrices exploiting BlueTooth traffic data. Proceedings of 91st Annual Meeting. Washington, D.C.: Transportation Research Board.

Bartin, B., Ozbay, K., Yanmaz-Tuzel, O. et al. 2006. Modeling and simulation of unconventional traffic circles. *Transportation Research Record*, 1965, 201–209.

Beegala, A., Hourdakis, J., and Michalopoulos, P.G. 2005. Methodology for performance optimization of ramp control strategies through microsimulation. *Transportation Research Record*, 1925, 87–98.

Bell, M.G.H. 1991a. The estimation of origin–destination matrices by constrained generalized least squares. *Transportation Research Part B*, 25(1), 13–22.

Bell, M.G.H. 1991b. The real time estimation of origin–destination flows in the presence of platoon dispersion. *Transportation Research Part B*, 25(2–3), 115–125.

Bella, F. 2005. Driving simulator validation for work-zone design. *Transportation Research Record*, 1937, 136–144.

Bella, F. 2008. Driving simulator for speed research on two-lane rural roads. *Accident Analysis and Prevention*, 40(3), 1078–1087.

Ben-Akiva, M., Bierlaire, M., Burton, D. et al. 2001. Network state estimation and prediction for real-time transportation management applications. *Networks and Spatial Economics*, 1(3/4), 291–318.

Ben-Akiva, M., Bierlaire, M., Koutsopoulos, H.N. et al. 2002. Real-time simulation of traffic demand–supply interactions within DynaMIT. In M. Gendreau and P. Marcotte, Eds., *Transportation and Network Analysis*. Amsterdam: Kluwer, pp. 19–36.

Benekohal, R.F. 1991. Procedure for validation of microscopic traffic flow simulation models. *Transportation Research Record*, 1320, 190–202.

Bertini, R.L. and Leal, M.T. 2005. Empirical study of traffic features at a freeway lane drop. *Journal of Transportation Engineering*, 131(6), 397–407.

Bierlaire, M. 2002. The total demand scale: a new measure of quality for static and dynamic origin–destination trip table. *Transportation Research Part B*, 36(9), 837–850.

Bierlaire, M. and Crittin, F. 2004. An efficient algorithm for real-time estimation and prediction of dynamic OD tables. *Operation Research*, 52(1), 116–127.

Blaauw, G. 1984. Driving experience and task demands in simulators and instrumented cars: a validation study. Amsterdam: Soesterberg: Institute for Perception.

Blana, E. 1996. A survey of driving research simulators around the world. Working Paper 481. University of Leeds.

Bloomberg, L. and Dale, J. 2000. A comparison of the VISSIM and CORSIM traffic simulation models on a congested network. *Transportation Research Record*, 1727, 52–60.

Borzacchiello, M.T., Steenbruggen, J., Nijkamp, P. et al. 2009. The use of data from mobile phone networks for transportation applications. Proceedings of 89th Annual Meeting. Washington, DC: Transportation Research Board.

Botma, H. 1999. The free speed distribution of drivers: estimation approaches. In *Five Years: Crossroads of Theory and Practice*. Delft University Press, pp. 1–22.

Bowman, J.L. and Ben-Akiva, M. 1998. Activity-based travel demand model systems. In P. Marcotte and S. Nguyen, Eds., *Equilibrium and Advanced Transportation Modeling*. Amsterdam: Kluwer, pp. 27–46.

Box, G., Hunter, J.S., and Hunter, W.G. 1978. *Statistics for Experimenters: Introduction to Design, Data Analysis, and Model Building*. Hoboken, NJ: Wiley.

Brackstone, M., Fisher, G., and McDonald, M. 2001. The use of probe vehicles on motor-ways: some empirical observations, Proceedings of 8th World Congress on Intelligent Transport Systems (CD-ROM).

Brenninger-Göthe, M., Jörnsten, K.O., and Lundgren, J.T. 1989. Estimation of origin–destination matrices from traffic counts using multi-objective programming formulations. *Transportation Research Part B*, 23(4), 257–269.

Briedis, P. and Samuels, S. 2010. The accuracy of inductive loop detectors. Presentation to ARRB Group, October 15, 2010. http://www.arrb.com.au/admin/file/content13/c6/Briedis,%20Paul,%20Accuracy%20of%20inductive%20loop%20detectors%20%28w%29.pdf

Brockfeld, E., Kuhne, R.D., and Wagner, P. 2004. Calibration and validation of microscopic traffic flow models. *Transportation Research Record*, 1876, 62–70.

Brockfeld E., Kuhne, R.D., and Wagner, P. 2005. Calibration and validation of microscopic models of traffic flow. *Transportation Research Record*, 1934, 179–187.

Bruno, F. and Greco, F. 2008. Clustering compositional data trajectories. CODAWORK '08.

Caliper, 2006. About Transmodeler. http://caliper.com/transmodeler/default.htm

Campolongo, F., Cariboni, J., and Saltelli, A. 2007. An effective screening design for sensitivity analysis of large models. *Environmental Modeling and Software*, 22(10), 1509–1518.

Carsten, O., Groeger, J., Blana, E. et al. 1997. *Driving Performance in EPSRC Driving Simulator* Leeds: LADS.

Casas, J., Perarnau, J., Garcia, J.L. et al. 2010. Trajectory data as a validation data for traffic models: NGSIM case. Proceedings of Traffic Flow Theory Summer Meeting and Conference, Annecy, France, July 7–9.

Cascetta, E. 1984. Estimation of origin-destination matrices from traffic counts and survey data: a generalized least squares estimator. *Transportation Research Part B*, 18(4–5), 289–299.

Cascetta, E. 2001. *Transportation Systems Engineering: Theory and Methods.* Amsterdam: Kluwer.

Cascetta, E., Inaudi, D., and Marquis, G. 1993. Dynamic estimators of origin–destination matrices using traffic counts. *Transportation Science*, 27, 363–373.

Cascetta, E. and Nguyen, S. 1988. A unified framework for estimating or updating origin-destination matrices from traffic counts. *Transportation Research Part B*, 22(6), 437–455.

Cascetta, E. and Postorino, M.N. 2001. Fixed point approaches to the estimation of O-D matrices using traffic counts on congested networks. *Transportation Science*, 35(2), 134–147.

Cassidy, M.J. and Mauch, M. 2001. An observed traffic pattern in long freeway queues. *Transportation Research Part A*, 35(2), 143–156.

Chang, T.H. and Li, Z.Y. 2002. Optimization of mainline traffic via an adaptive coordinated ramp metering control model with dynamic OD estimation. *Transportation Research Part C*, 10(2), 99–120.

Chen, A., Yang, H., Lo, H.K. et al. 2002. Capacity reliability of a road network: an assessment methodology and numerical results. *Transportation Research Part B*, 36(3), 225–252.

Chen, C.H., Schonfeld, P., and Paracha, J. 2005. Work zone optimization for two-lane highway resurfacing projects with an alternate route. *Transportation Research Record*, 1911, 51–66.

Chen, M. and Chien, S. 2001. Dynamic freeway travel time prediction using probe vehicle data: link-based versus path-based. Proceedings of 80th Annual Meeting. Washington, DC: Transportation Research Board.

Cheng, P., Qiu, Z., and Ran, B. 2006. Particle filter-based traffic state estimation using cell phone network data. Proceedings of IEEE ITSC.

Chi, S. and Caldas, C.H. 2011. Automated object identification using optical video cameras on construction sites. *Computer-Aided Civil and Infrastructure Engineering*, 26(5), 368–380.

Chiabaut, N. and Leclercq, L. 2011. Wave velocity estimation through cumulative vehicle count curves automatic analysis. *Transportation Research Record*, 2249, 1–6.

Chiabaut, N., Leclercq, L., and Buisson, C. 2010. From heterogeneous drivers to macroscopic patterns in congestion. *Transportation Research Part B*, 44(2), 299–308.

Chipperfield, A., Fleming, P., Pohlheim, H. et al. 2010. Genetic Algorithm Toolbox User's Guide. http://www.shef.ac.uk/acse/research/ecrg/gat.html

Chiu, Y.C., Zhou, L., and Song, H. 2010. Development and calibration of the anisotropic mesoscopic simulation model for uninterrupted flow facilities. *Transportation Research Part B*, 44(1), 152–174.

Cho, Y. and Rice, J. 2006. Estimating velocity fields on a freeway from low-resolution videos. *IEEE Transactions on Intelligent Transportation Systems*, 7(4), 463–469.

Chong, L., Abbas, M., Flintsch, A. et al. 2013. A rule-based neural network approach to model driver naturalistic behavior in traffic. *Transportation Research Part C*, 32, 207–223.

Chu, L., Liu, H.X., Oh, J. et al. 2004. A calibration procedure for microscopic traffic simulation. Proceedings of 83rd Annual Meeting. Washington, DC: Transportation Research Board.

Chu, L., Oh, J.S., and Recker, W. 2005. Adaptive Kalman filter-based freeway travel time estimation. Proceedings of 84th Annual Meeting. Washington, DC: Transportation Research Board.

Chung, E., Sarvi, M., Murakami, Y. et al. 2003. Cleansing of probe car data to determine trip OD, Proceedings of 21st ARRB and 11th REAAA Conference (CD-ROM).

Ciuffo, B. and Punzo, V. 2010a. Verification of traffic micro-simulation model calibration procedures: analysis of goodness-of-fit measures. Proceedings of 89th Annual Meeting. Washington, DC: Transportation Research Board.

Ciuffo, B. and Punzo, V. 2010b. Kriging meta-modeling in the verification of traffic micro-simulation calibration procedure: optimization algorithms and goodness of fit measures. Proceedings of Traffic Flow Theory Summer Meeting and Conference, Annecy, France, July 7–9.

Ciuffo, B., V. Punzo and M. Montanino, 2014. Global sensitivity analysis techniques to simplify the calibration of traffic simulation models. Methodology and application to the IDM car-following model. IET Intelligent Transportation Systems (in press). DOI: 10.1109/TITS.2013.2287720

Ciuffo, B., Punzo, V., and Montanino, M. 2012. The Calibration of Traffic Simulation Models: Report on the Assessment of Different Goodness-of-Fit Measures and Optimization Algorithms–COST Action TU0903 (MULTITUDE). JRC Scientific and Technical Reports, JRC 68403, Publications Office of the European Union, Luxembourg.

Ciuffo, B., Punzo, V., and Quaglietta, E. 2010. Verification of traffic micro-simulation model calibration procedures: analysis of goodness–of-fit measures. Proceedings of 89th Annual Meeting. Washington, DC: Transportation Research Board.

Ciuffo, B., Punzo, V., and Torrieri, V. 2007. A framework for calibrating a microscopic simulation model. Proceedings of 86th Annual Meeting. Washington, DC: Transportation Research Board.

Ciuffo, B., Punzo, V. and Torrieri, V. 2008. A comparison between simulation-based and model-based calibrations of traffic flow micro-simulation models. *Transportation Research Record*, 2088, 36–44.

Coifman, B. 2002. Estimating travel time and vehicle trajectories on freeways using dual loop detectors. *Transportation Research Part A*, 36(4), 351–364.

Coifman, B. and Cassidy, M. 2002. Vehicle re-identification and travel time measurement on congested freeways. *Transportation Research Part A*, 36(10), 899–917.

Coifman, B. and Kim, S. 2008. Speed estimation and length based vehicle classification from freeway single loop detectors. Proceedings of 88th Annual Meeting. Washington: Transportation Research Board.

Coifman, B. and Kim, S. 2009. Speed estimation and length based vehicle classification from freeway single-loop detectors. *Transportation Research Part C*, 17(4), 349–364.

Coifman, B. and Wang, Y. 2005. Average velocity of waves propagating through congested freeway traffic. In *Transportation and Traffic Theory: Flow, Dynamics, and Human Interaction*. 16th International Symposium on Transportation and Traffic Theory.

Coifman, B. 1999. Vehicle re-identification and travel time measurement using loop detector speed traps. PhD thesis. Berkeley: University of California, ITS-DS-98-2.

Coifman, B. 2001. Improved velocity estimation using single loop detectors. *Transportation Research Part A*, 35(10), 863–880.

Coifman, B., Beymer, D., McLauchlan, P. et al. 1998. A real-time computer vision system for vehicle tracking and traffic surveillance. *Transportation Research Part C*, 6(4), 271–288.

Conn, A.R., Gould, N.I.M., and Toint, P.L. 1991. A globally convergent augmented Lagrangian algorithm for optimization with general constraints and simple bounds. *SIAM Journal on Numerical Analysis*, 28(2), 545–572.

Conn, A.R., Gould, N.I.M., and Toint P.L. 1997. A globally convergent augmented Lagrangian barrier algorithm for optimization with general inequality constraints and simple bounds. Mathematics of Computation, 66(217), pp. 261-288.

COST TU0702, 2011. What is COST TU0702 all about? http://www.cost-tu0702.org/

COST TU0903, 2013. Welcome to MULTITUDE. http://www.multitude-project.eu/. Last accessed October 2013.

Cremer, M. and Papageorgiou, M. 1981. Parameter identification for a traffic flow model. *Automatica*, 17, 837–843.

Cuevas, A., Febrero, M., and Fraiman, R. 2003. An ANOVA test for functional data. *Computational Statistics and Data Analysis*, 47(1), 111–122.

Cukier, R.I., Fortuin, C.M., Schuler, K.E. et al. 1973. Study of the sensitivity of coupled reaction systems to uncertainties in rate coefficients. *Journal of Chemical Physics*, 26, 1–42.

Daganzo, C. 1994. The cell transmission model: a dynamic representation of highway traffic consistent with the hydrodynamic theory. *Transportation Research Part B*, 28(4), 269–287.

Dahlkamp, H., Kaehler, A., Stavens, D. et al. 2006. Self-supervised monocular road detection in desert terrain. Proceedings of Robotics: Science and Systems.

Dailey, D. 1999. A statistical algorithm for estimating speed from single loop volume and occupancy measurements. *Transportation Research Part B*, 33(5), 313–322.

Da Veiga, S., Wahl, F., and Gamboa, F. 2009. Local polynomial estimation for sensitivity analysis on models with correlated inputs. *Tachnometrics*, 51(4), 452–463.

De Rocquigny, E., Devictor, N., and Tarantola, S. 2008. *Uncertainty in Industrial Practice: A Guide to Quantitative Uncertainty Management.* New York: Wiley.

Desmond, P.A. and Matthews, G. 1997. Implication of task-induced fatigue effects for in-vehicle countermeasures to driver fatigue. *Accident Analysis and Prevention,* 29(4), 515–523.

Doan, D.L., Ziliaskopoulos, A., and Mahmassani, H. 1999. On-line monitoring system for real-time traffic management applications. *Transportation Research Record,* 1678, 142–149.

Dowling, R., Skabardonis, A., Halkias, J. et al. 2004. Guidelines for calibration of microsimulation models: framework and applications. *Transportation Research Record,* 1876, 1–9.

DRIVE C2X.2011. Preparation of driving implementation and evaluation of C2X communication technology, http://www.pre-drive-c2x.eu/

Drud, A.S. 1994. CONOPT: a large-scale GRG code. *ORSA Journal of Computing,* 6(2), 207–216.

Edie, L. 1965. Discussion of traffic stream measurements and definitions. In *Proceedings of 2nd International Symposium on Theory of Traffic Flow,* Paris, pp. 139–154.

Ehlert, A., Bell, M.G.H., and Grosso, S. 2006. The optimization of traffic count locations in road network. *Transportation Research Part B,* 40(6), 460–479.

El Faouzi, N.E., de Mouzon, O., and Billot, R. 2010. Assessing individual driver behaviour and macroscopic traffic changes due to the weather conditions. *Advances in Transportation Studies,* 21, 33–46.

Ellingrod, V.L., Perry, P.J., Yates, R.W. et al. 1997. The effects of anabolic steroids on driving performance as assessed by the Iowa Driver Simulator. *American Journal of Drug and Alcohol Abuse,* 23(4), 623–636.

Erlander, S. 1971. A mathematical model for traffic on a two-lane road with some empirical results. *Transportation Research,* 5(2), 135–147.

Ervin, R.D., MacAdam, C.C., Gilbert, K. et al. 1991. Quantitative characterization of the vehicle motion environment. In *Proceedings of 2nd Vehicle Navigation and Information Systems Conference,* pp. 1011–1029.

Farah, H., Bekhor, S., and Polus, A. 2009. Risk evaluation by modeling of passing behavior on two-lane rural highways. *Accident Analysis and Prevention,* 41(4), 887–894.

Farah, H., Polus, A., Bekhor, S. et al. 2009. A passing gap acceptance model for two-lane rural highways. *Transportmetrica,* 5(3), 159–172.

Farah, H. and Toledo, T. 2010. Passing behavior on two-lane highways. *Transportation Research Part F,* 13(6), 355–364.

Fasano, G. and Franceschini, A. 1987. A multidimensional version of the Kolmogorov-Smirnov test. *Monthly Notices of Royal Astronomical Society,* 225, 155–170.

Fellendorf, M. and Vortisch, P. 2001. Validation of the microscopic traffic flow model VISSIM in different real-world situations. Proceedings of 80th Annual Meeting. Washington, DC: Transportation Research Board.

FHWA, 2001. Heavy Vehicle Travel Information System Field Manual. https://www.fhwa.dot.gov/ohim/tvtw/hvtis.pdf

FHWA, 2005. CORSIM website. http://ops.fhwa.dot.gov/trafficanalysistools/corsim.htm

Fishman, G.S. 1978. *Principles of Discrete Event Simulation*. New York: Wiley.

Fisk, C. 1988. On combining maximum entropy trip matrix estimation with user equilibrium. *Transportation Research Part B*, 22(1), 69–73.

Flexman, R.E. and Stark, E.A. 1987. In *Training Simulators: Handbook of Human Factors*, G. Salvendy, Ed. New York: Wiley, pp. 1012–1038.

Florian, M. and Chen, Y. 1992. A successive linear approximation method for the OD matrix adjustment problem. Publication 807, Transportation Research Centre, Montreal University.

Florian, M. and Chen, Y. 1995. A coordinate descent method for the bilevel OD matrix adjustment problem. *International Transactions in Operational Research*, 2(2), 165–179.

Flötteröd, G. 2009. Cadyts: a free calibration tool for dynamic traffic simulations. Proceedings of Swiss Transport Research Conference, Ascona.

Flötteröd, G. and Bierlaire, M. 2009. Improved estimation of travel demand from traffic counts by a new linearization of the network loading map. Proceedings of European Transport Conference, Noordwijkerhout, Netherlands.

Flötteröd, G. M. Bierlaire, and K. Nagel. Bayesian demand calibration for dynamic traffic simulations. Transportation Science, 45(4):541–561, 2011.

Flötteröd, G. Y. Chen, and K. Nagel. Behavioral calibration and analysis of a large-scale travel microsimulation. Networks and Spatial Economics, 12(4):481–502, 2012.

Fraser, D.A., Hawken, R.E., and Warnes, A.M. 1994. Effects of extra signals on drivers' distance keeping: a simulation study. *IEEE Transactions on Vehicular Technology*, 43(4), 1118–1124.

Frey, H.C. and Patil, S.R. 2002. Identification and review of sensitivity analysis methods. *Risk Analysis*, 22(3), 553–578.

Gallant, A.R. and Goebel, J.J. 1976. Nonlinear regression with auto-correlated errors. *Journal of American Statistical Association*, 71, 961–967.

Gazis, D.C. and Knapp, C.H. 1971. On-line estimation of traffic densities from time series of flow and speed. *Transportation Science*, 5, 283–301.

Ghosh, D. and Knapp, C.H. 1978. Estimation of traffic variables using a linear model of traffic flow. *Transportation Research*, 12(6), 395–402.

Gipps, P.G. 1981. A behavioural car-following model for computer simulation. *Transportation Research Part B*, 15, 105–111.

Glover, F. 1998. A template for scatter search and path relinking in artificial evolution. In J.K. Hao et al., Eds., *Artificial Evolution*. Lecture Notes in Computer Science 1363. Heidelberg: Springer, pp. 13–54.

Google Maps. 2012. http://maps.google.com/

Grant, C., Gillis, B., and Guensler, R. 1999. Collection of Vehicle Activity Data by Video Detection for Use in Transportation Planning: Research Report. Embry-Riddle Aeronautical University, Daytona Beach, FL.

Graves, T., Karr, A., Rouphail, N. et al. 1998. Real-Time Prediction of Incipient Congestion on Freeways from Detector Data: Technical Report. National Institute of Statistical Sciences, Research Triangle Park, NC.

Greenberg, J. 2004. Physical fidelity and driving simulation. Proceedings of 83rd Annual Meeting. Washington: Transportation Research Board. www.engineering. uiowa.edu/simusers/Archives/TRB 2004/Greenberg.ppt

Gupta, A. 2005. Observability of origin–destination matrices for dynamic traffic assignment. Master's thesis. Cambridge: Massachusetts Institute of Technology.

Haan, C.T., Storm, D.E., Al-Issa, T. et al. 1998. Effect of parameter distribution on uncertainty analysis of hydrologic models. *Transactions of ASAE*, 41(1), 65–70.

Hall, F., Bloomberg, L., Rouphail, N.M., et al. 2000. Validation results for four models of oversaturated freeway facilities. *Transportation Research Record*, 1710, 161–170.

Hamdar, S.H. and Mahmassani, H.S. 2008. Driver car-following behavior: from a discrete event process to a continuous set of episodes. Proceedings of 87th Annual Meeting. Washington, DC: Transportation Research Board.

Harik, G.R., Lobo, F.G., and Goldberg, D.E. 1999. The compact genetic algorithm. *IEEE Transactions on Evolutionary Computation*, 3(4), 287–297.

Hazelton, M.L. 2000. Estimation of origin–destination matrices from link flows on uncongested networks. *Transportation Research Part B*, 34(7), 549–566.

Hazelton, M.L. 2003. Some comments on origin–destination matrix estimation. *Transportation Research Part A*, 37(10), 811–822.

He, R.R., Kornhauser, A.L., and Ran, B. 2001. Estimation of time-dependent OD demand and route choice from link flows. Princeton, NJ: Department of Operations Research and Financial Engineering, Princeton University.

Heidemann, J., Silva, F., Wang, X. et al. 2008. SURE-SE-Sensors for Unplanned Roadway Events: Simulation and Evaluation. Draft Final Report, METRANS Project 04-08.

Helbing, D. and Tilch, B. 1998. Generalized force model of traffic dynamics. *Physical Reviews E*, 58(1), 133–138.

Helly, W. 1959. Simulation of bottlenecks in single lane traffic flow. In *Proceedings of Symposium on Theory of Traffic Flow*, pp. 207–238.

Helton, J.C. 1993. Uncertainty and sensitivity analysis techniques for use in performance assessment for radioactive waste disposal. *Reliability Engineering and System Safety*, 42(2–3), 327–367.

Herrera, J.C. and Bayen, A.M. 2007. Traffic flow reconstruction using mobile sensors and loop detector data. Proceedings of 87th Annual Meeting. Washington, DC: Transportation Research Board.

Hidas, P. and Wagner, P. 2004. Review of data collection methods for microscopic traffic simulation. 10th World Conference on Transport Research.

Highway Agency. 1996. *Design Manual for Roads and Bridges*, Vol. 12: Traffic Appraisal of Road Schemes, Traffic Appraisal in Urban Areas. London: Her Majesty's Stationery Office.

Highway Agency. 2012. *Live Traffic Information Covering England's Motorways and Major Roads*. London: Her Majesty's Stationery Office.

Highway Research Board. 2000. *Highway Capacity Manual: Technical Report*.

Hoban, C.J. 1980. Overtaking lanes on two-lane rural highways. PhD Thesis. Melbourne: Monash University.

Høeg, P. 1995. *Borderlines*. New York: Random House.

Holland, J.H. 1975. *Adaptation in Natural and Artificial Systems*. Ann Arbor: University of Michigan Press.

Hollander, Y. and Liu, R. 2005. Calibration of a traffic microsimulation model as a tool for estimating the level of travel time variability. 10th meeting of EURO Working Group on Transportation, Poznan, Poland.

Hollander, Y. and Liu, R. 2008a. The principles of calibrating traffic microsimulation models. *Transportation*, 35(3), 347–362.

Hollander, Y. and Liu, R. 2008b. Estimation of the distribution of travel times by repeated simulation. *Transportation Research Part C*, 16(2), 212–231.

Holm, S. 1979. A simple sequentially rejective multiple test procedure. *Scandinavian Journal of Statistics*, 6(2), 65–70.

Homma, T. and Saltelli, A. 1996. Importance measures in global sensitivity analysis of nonlinear models. *Reliability Engineering and System Safety*, 52(1), 1–17.

Hoogendoorn, S.P. and Hoogendoorn, R.G. 2010a. Calibration of microscopic traffic flow models using multiple data sources. *Philosophical Transactions of Royal Society A*, 368, 4497–4517.

Hoogendoorn, S.P. and Hoogendoorn, R.G. 2010b. Generic calibration framework for joint estimation of car-following models by using microscopic data. *Transportation Research Record*, 2188, 37–45.

Hoogendoorn, R.G., Hoogendoorn, S.P., Brookhuis, K.A. et al. 2010. Mental workload, longitudinal driving behavior and adequacy of car-following models for incidents in the other driving lane. *Transportation Research Record*, 2188, 64–73.

Hoogendoorn, R.G., Hoogendoorn, S.P., and Brookhuis, K.A. 2011a. Empirical adaptation effects and modeling of longitudinal driving behavior in case of emergency situations. Conference on Traffic and Granular Flow, Moscow.

Hoogendoorn, R.G., Hoogendoorn, S.P., Brookhuis, K.A., et al. 2011b. Adaptation effects in longitudinal driving behavior: mental workload and psycho-spacing models in case of fog. *Transportation Research Record*, 2249, 20–28.

Hoogendoorn, S.P. and Van Lint, H. 2010. Contribution to CIE4831. Delft University of Technology.

Hoogendoorn, S.P., van Zuylen, H.J., Schreuder, M. et al. 2003. Microscopic traffic data collection by remote sensing. *Transportation Research Record*, 1885, 121–128.

Horowitz, A.J. 2000. *Quick Response System II, Version 6 Reference Manual*. Center for Urban Transportation Studies, University of Wisconsin and AJH Associates.

Hough, P. 1962. Method and Means for Recognizing Complex Patterns. U.S. Patent 3,069,654.

Hourdakis, J., Michalopoulos, P.G., and Kottommannil, J. 2003. Practical procedure for calibrating microscopic traffic simulation models. *Transportation Research Record*, 1852, 130–139.

Hu, S.R., Peeta, S., and Chu, C.H. 2009. Identification of vehicle sensor locations for link-based network traffic applications. *Transportation Research Part B*, 43(8–9), 873–894.

Huynh, N., Mahmassani, H., and Tavana, H. 2002. Adaptive speed estimation using transfer function models for real-time dynamic traffic assignment operation. Proceedings of 81st Annual Meeting. Washington: Transportation Research Board.

INRO. 2006. About EMME/2. http://www.inro.ca/en/products/emme2/index.php

Iooss, B. 2010. Metamodeling with Gaussian processes. Presented at 6th Summer School on Sensitivity Analysis, Florence, Italy, September 14–17.

IPCC. 2000. Good Practice Guidance and Uncertainty Management in National Greenhouse Gas Inventories. http://www.ipcc-nggip.iges.or.jp/public/gp/english/

Isukapalli, S.S. 1999. Uncertainty analysis of transport transformation models. PhD thesis. New Brunswick: Rutgers State University of New Jersey.

Jacques, J., Lavergne, C., and Devictor, N. 2006. Analysis in presence of model uncertainties and correlated inputs. *Reliability Engineering and System Safety*, 91(10–11), 1126–1134.

Jain, M. and Coifman, B. 2005. Improved speed estimates from freeway traffic detectors. *ASCE Journal of Transportation Engineering*, 131(7), 483–495.

Jayakrishnan, R., Oh, J., and Sahraoui, A.K. 2001. Calibration and path dynamics issues in microscopic simulation for advanced traffic management and information systems. *Transportation Research Record*, 1771, 9-17.

Jha, M., Gopalan, G., Garms, A., et al. 2004. Development and calibration of a large-scale microscopic traffic simulation model. *Transportation Research Record*, 1876, 121–131.

Ji, X. and Prevedouros, P.D. 2005a. Comparison of methods for sensitivity and uncertainty analysis of signalized intersections analyzed with the HCM. *Transportation Research Record*, 1920, 56-64.

Ji, X. and Prevedouros, P.D. 2005b. Effects of parameter distributions and correlations on uncertainty analysis of HCM delay model. *Transportation Research Record*, 1920, 118–124.

Ji, X. and Prevedouros, P.D. 2006. Probabilistic analysis of highway capacity manual delay for signalized intersections. *Transportation Research Record*, 1988, 67–75.

Ji, X. and Prevedouros, P.D. 2007. Probabilistic analysis of HCM models: case study of signalized intersections. *Transportation Research Record*, 2027, 58–64.

Johnson, R.A. and Wichern, D.W. 1998. *Applied Multivariate Statistical Analysis*, 4th ed. New York: Prentice Hall.

Kaplan, E. and Meier, P. 1958. Non-parametric estimation for incomplete observations. *Journal of American Statistical Association*, 53(282), 457–481.

Kaptein, N.A., Theeuwes, J., and Van der Horst, A.R.A. 1996. Driving simulator validity: some considerations. *Transportation Research Record*, 1550, 30–36.

Kaptein, N.A., Van der Horst, R., and Hoekstra, W. 1996. Visual Information in a Driving Simulator: A Study on Braking Behaviour and Time to Collision. Soesterberg, Netherlands: TNO Human Factor Research Institute.

Kastrinaki, V., Zervakis, M., and Kalaitzakis, K. 2003. A survey of video processing techniques for traffic applications. *Image and Vision Computing*, 21(4), 359–381.

Kerner, B.S. 1998. Experimental features of self-organization in traffic flow. *Physical Review Letters*, 81(17), 3797–3800.

Kerner, B.S. and Rehborn, H. 1997. Experimental properties of phase transitions in traffic flow. *Physical Review Letters*, 79(20), 4030–4033.

Kesting, A. and Treiber, M. 2008a. Calculating travel times from reconstructed spatiotemporal traffic data. Proceedings of 4th International Symposium on Networks for Mobility, Stuttgart. Kesting, A. and Treiber, M. 2008b. Calibrating car-following models by using trajectory data. *Transportation Research Record*, 2088, 148–156.

Kiefer, J. and Wolfowitz, J. 1952. Stochastic estimation of the maximum of a regression function. *Annals of Mathematical Statistics*, 23(3), 462–466.

Kim, H., Baek, S., and Lim, Y. 2001. Origin–destination matrices estimated with a genetic algorithm from link traffic counts. *Transportation Research Record*, 1771, 156–163.

Kim, K.O. and Rilett, L.R. 2003. Simplex-based calibration of traffic microsimulation models with intelligent transportation systems data. *Transportation Research Record*, 1855, 80–89.

Kim, K.O. and Rilett, L.R. 2004. A genetic algorithm-based approach to traffic microsimulation calibration using ITS data. Proceedings of 83rd Annual Meeting. Washington, : Transportation Research Board.

Kim, S.J., Kim, W., and Rilett, L.R. 2005. Calibration of microsimulation models using nonparametric statistical techniques. *Transportation Research Record*, 1935, 111–119.

Kim, J.W. and Mahmassani, H.S. 2011. Correlated parameters in driving behavior models: car-following example and implications for traffic microsimulation. *Transportation Research Record*, 2249, 62–77.

Kirkpatrick, S., Gelatt, C.D., and Vecchi, M.P. 1983. Optimization by simulated annealing. *Science*, 220(4598), 671–680.

Kleijnen, J.P.C. 1995. Verification and validation of simulation models. *European Journal of Operational Research*, 82, 145–162.

Kleijnen, J.P.C. 1999. Validation of models: statistical techniques and data availability. Paper presented at 1999 Winter Simulation Conference, Phoenix, AZ.

Knoop, V.L., Buisson, C., Wilson, E., et al. 2011. Number of lane changes determined by splashover effects in loop detector counts. Proceedings of 90th Annual Meeting. Washington, DC: Transportation Research Board.

Knoop, V.L., Hoogendoorn, S.P., and Van Zuylen, H.J. 2009a. Empirical differences between time mean speed and space mean speed. *Traffic and Granular Flow*, 7, 351–356.

Knoop, V.L., Hoogendoorn, S.P., and Van Zuylen, H.J. 2009b. Processing traffic data collected by remote sensing. *Transportation Research Record*, 2129, 55–63.

Kohan, R.R. and Bortoff, S.A. 1998. An observer for highway traffic systems. In *Proceedings of 37th IEEE Conference on Decision and Control*, pp. 1012–1017.

Kondyli, A and Elefteriadou, L. 2010. Driver behavior at freeway-ramp merging areas based on instrumented vehicle observations. Proceedings of 89th Annual Meeting. Washington, : Transportation Research Board.

Koutsopoulos, H.N. and Habbal, M. 1994. Effect of intersection delay modeling on the performance of traffic equilibrium models. *Transportation Research Part A*, 28(2), 133–149.

Kunde, K.K. 2002. Calibration of mesoscopic traffic simulation models for dynamic traffic assignment. Master's thesis. Cambridge: Massachusetts Institute of Technology.

Kurkjian, A., Gershwin, S.B., Houpt, P.K., et al. 1980. Estimation of roadway traffic density on freeways using presence detector data. *Transportation Science*, 14(3), 232–261.

Lagarias, J.C., Reeds, J.A., Wright, M.H., et al. 1998. Convergence properties of the Nelder-Mead simplex method in low dimensions. *SIAM Journal of Optimization*, 9(1), 112–147.

Lam, W.H.K. and Lo, H.P. 1991. Estimation of origin-destination matrix from traffic counts: a comparison of entropy maximizing and information minimizing models. *Transportation Planning and Technology*, 16(2), 85–104.

Law, A.M. 2007. *Simulation Modeling and Analysis*, 4th ed. New York: McGraw Hill.

Law, A.M. and Kelton, W.D. 2000. *Simulation Modeling and Analysis*, 3rd ed. New York: McGraw Hill.

Lawe, S., Lobb, J., Sadek, A.W., et al. 2009. TRANSIMS implementation in Chittenden County, Vermont: development, calibration, and preliminary sensitivity analysis. *Transportation Research Record*, 2132, 113–121.

Leclercq, L. 2005. Calibration of flow density relationships in urban streets. Proceedings of 84th Annual Meeting. Washington, DC: Transportation Research Board.

Lee, D.H., Yang, X., and Chandrasekar, P. 2001. Parameter calibration for PARAMICS using genetic algorithm. Proceedings of 80th Annual Meeting. Washington: Transportation Research Board.

Lee, H. and Coifman, B. 2010. Algorithm to identify splashover errors at freeway loop detectors. Proceedings of 89th Annual Meeting. Washington, DC: Transportation Research Board.

Lee, J.B. and Ozbay, K. 2009. A new calibration methodology for microscopic traffic simulation using enhanced simultaneous perturbation stochastic approximation (E-SPSA) Approach. *Transportation Research Record*, 2124, 233–240.

Lenhart, D., Hinz, S., Leitloff, J. et al. 2008. Automatic traffic monitoring based on aerial image sequences. *Pattern Recognition and Image Analysis*, 18(3), 400–405.

Leung, S. and Starmer, G. 2005. Gap acceptance and risk-taking by young and mature drivers, both sober and alcohol intoxicated, in a simulated driving task. *Accident Analysis and Prevention*, 37(6), 1056–1065.

Leurent, F. 1998. Sensitivity and error analysis of the dual criteria traffic assignment model. *Transportation Research Part B*, 32(3), 189–284.

Leutzbach, W. 1987. *Introduction to the Theory of Traffic Flow*. Berlin: Springer.

Li, Z., Liu, H., and Zhang, K. 2009. Sensitivity analysis of PARAMICS based on 2K-P fractional factorial design. In *Proceedings of 2nd International Conference on Transportation Engineering*, pp. 3633–3638.

Lighthill, M. and Whitham, G. 1955. On kinematic waves II: theory of traffic flow on long crowded roads. In *Proceedings of Royal Society of London A*, 229(1178), 317–345.

Lindo Systems, Inc. 2002. *API User Manual 2.0*. Chicago.

Lindveld, C. and Thijs, R. 1999. On-line travel time estimation using inductive loop data: the effect of instrumentation peculiarities on travel time estimation quality. Proceedings of 6th ITS World Congress, Toronto.

Lindveld, K. 2003. Dynamic OD matrix estimation: a behavioural approach. PhD thesis. Delft University of Technology.

Liu, R., Van Vliet, D., and Watling, D.P. 1995. DRACULA: Dynamic route assignment combining user learning and microsimulation. Proceedings of PTRC Summer Annual Conference.

Liu, S.S. and Fricker, J.D. 1996. Estimation of a trip table and the theta parameter in a stochastic network. *Transportation Research Part A*, 30(4), 287–305.

Lo, H.P. and Chan, C.P. 2003. Simultaneous estimation of an origin–destination matrix and link choice proportions using traffic counts. *Transportation Research Part A*, 37(9), 771–788.

Lo, H.P., Zhang, N., and Lam, W.H.K. 1996. Estimation of an origin–destination matrix with random link choice proportions: a statistical approach. *Transportation Research Part B*, 30(4), 309–324.

Lo, H.P., Zhang, N., and Lam, W.H.K. 1999. Decomposition algorithm for statistical estimation of OD matrix with random link choice proportions from traffic counts. *Transportation Research Part B*, 33(5), 369–385.

Lownes, N.E. and Machemehl, R.B. 2006. Sensitivity of simulated capacity to modification of VISSIM driver behavior parameters. *Transportation Research Record*, 1988, 102–110.

Lownes, N.E. and Machemehl, R.B. 2006. VISSIM: a multi-parameter sensitivity analysis. In Proceedings of Winter Simulation Conference, Article 4117765, pp. 1406–1413.

Lu, J., Rechtorik, M., and Yang, S. 1997. Analysis of AVI technology applications to toll collection services. *Transportation Research Record*, 1588, 18–25.

Lu, X. and Skabardonis, A. 2007. Freeway traffic shockwave analysis: exploring the NGSIM trajectory data. Proceedings of 86th Annual Meeting. Washington, DC: Transportation Research Board.

Lundgren, J.T. and Peterson, A. 2008. A heuristic for the bilevel origin–destination-matrix estimation problem. *Transportation Research Part B*, 42(4), 339–354.

Lundgren, J.T., Peterson, A., and Rydergren, C. 2006. A Heuristic for the Estimation of Time-Dependent Origin–Destination Matrices from Traffic Counts, SE-60174. Linköping University, Norrköping, Sweden.

Ma, J., Dong, H., and Zhang, H.M. 2007. Calibration of microsimulation with heuristic optimization methods. *Transportation Research Record*, 1999, 208–217.

Ma, T. and Abdulhai, B. 2002. Genetic algorithm-based optimization approach and generic tool for calibrating traffic microscopic simulation parameters. *Transportation Research Record*, 1800, 6–15.

Ma, X. and Andreasson, I. 2006. Statistical analysis of driver behavior data in different regimes of the car-following stage. *Transportation Research Record*, 2018, 87–96.

Mahanti, B.P. 2004. Aggregate calibration of microscopic traffic simulation models. Master's thesis. Cambridge: Massachusetts Institute of Technology.

Maher, M. 1983. Inferences on trip matrices from observations on link volumes: a Bayesian statistical approach. *Transportation Research Part B*, 17(6), 435–447.

Maher, M., Zhang, X., and Van Vliet, D. 2001. A bilevel programming approach for trip matrix estimation and traffic control problems with stochastic user equilibrium link flows. *Transportation Research Part B*, 35(1), 23–40.

Maher, M.J. and Zhang, X. 1999. Algorithms for the solution of the congested trip matrix estimation problem. In *Proceedings of 14th International Symposium on Transportation and Traffic Theory*, pp. 445–469.

Mahmassani, H.S. 2002. Dynamic network traffic assignment and simulation methodology for advanced system management applications. Proceedings of 81st Annual Meeting. Washington: Transportation Research Board.

Mahut, M., Florian, M., Tremblay, N. et al. 2004. Calibration and application of a simulation-based dynamic traffic assignment model. *Transportation Research Record*, 1876, 101–111.

Malinovskiy, Y., Zheng, J., and Wang, Y. 2009. Model-free video detection and tracking of pedestrians and bicyclists. *Computer-Aided Civil and Infrastructure Engineering*, 24(3), 157–168.

Maryak, J.L. and Chin, D.C. 2008. Global random optimization by simultaneous perturbation stochastic approximation. *IEEE Transactions on Automatic Control*, 53(3), 780–783.

Marzano, V., Papola, A., and Simonelli, F. 2009. Limits and perspectives of effective OD matrix correction using traffic counts. *Transportation Research Part C*, 17(2), 120–132.

Mathew, T.V. and Radhakrishnan, P. 2010. Calibration of microsimulation models for non-lane-based heterogeneous traffic at signalized intersections. *Journal of Urban Planning and Development*, 136(1), 59–66.

McKinnon, K.I.M. 1996. Convergence of Nelder–Mead simplex method to a non-stationary point. *SIAM Journal of Optimization*, 9, 148–158.

McSharry, P.E. and Smith, L.A. 1999. Better non-linear models from noisy data: attractors with maximum likelihood. *Physical Review Letters*, 83, 4285–4288.

Meier, J. and Wehlan, H. 2001. Section-wise modeling of traffic flow and its application in traffic state estimation. In *Proceedings of Conference on Intelligent Transport Systems*, pp. 440–445.

Menneni, S., Sun, C., and Vortisch, P. 2008. Microsimulation calibration using speed flow relationships. *Transportation Research Record*, 2088, 1–9.

Merritt, E. 2004. Calibration and validation of CORSIM for Swedish road traffic conditions. Proceedings of 83rd Annual Meeting. Washington, DC: Transportation Research Board.

Messmer, A. and Papageorgiou, M. 2001. Freeway network simulation and dynamic traffic assignment with METANET tools. *Transportation Research Record*, 1776, 178–188.

Mihaylova, L., Boel, R., and Hegyi, A. 2007. Freeway traffic estimation within particle filtering framework. *Automatica*, 43(2), 290–300.

Mimbela, L.Y. and Klein, L.A. 2000. Summary of Vehicle Detection and Surveillance Technologies Used in Intelligent Transportation Systems. Report of Federal Highway Administration Intelligent Transportation System Joint Program Office. http://www.fhwa.dot.gov/ohim/tvtw/vdstits.pdf

Minge, E., Kotzenmacher, J., and Peterson, S. 2010. Evaluation of Non-Intrusive Technologies for Traffic Detection. Technical Report by SRF Consulting Group on behalf of Minnesota Department of Transportation. http://www.lrrb.org/PDF/201036.pdf

Minguez, R., Sanchez, P., Cambronero, S. et al. 2010. Optimal traffic plate scanning location for OD trip matrix and route estimation in road networks. *Transportation Research Part B*, 44(2), 282–298.

Molino, J., Katz, B., Duke, D., et al. 2005. Validate first; simulate later: a new approach used at the FHWA highway driving simulator. Proceedings of Driving Simulation Conference, Orlando, FL.

Morgan, M.G. and Henrion, M. 1990. *Uncertainty: A Guide to Dealing with Uncertainty in Quantitative Risk and Policy Analysis*. Cambridge, UK: Cambridge University Press.

Munoz, L., Sun, X., Sun, D., et al. 2004. Methodological calibration of the cell transmission model. In *Proceedings of Annual Control Conference*, pp. 798–803.

K. Nagel and Flötteröd G. 2012. Agent-based traffic assignment: going from trips to behavioral travelers. In C. Bhat, R.M. Pendyala, editors, Travel Behaviour Research in an Evolving World, pages 261–293, Emerald Group Publishing.

Nahi, N.E. and Trivedi, A.N. 1973. Recursive estimation of traffic variables: Section density and average speed. *Transportation Science*, 7(3), 269–286.

Nam, D.H. and Drew, D.R. 1996. Traffic dynamics: method for estimating freeway travel times in real time from flow measurements. *ASCE Journal of Transportation Engineering*, 122(3), 185–191.

Nanni, M. and Pedreschi, D. 2006. Time-focused clustering of trajectories of moving objects. *Journal of Intelligent Information Systems*, 27(3), 267–289.

Neimer, J. and Mohellebi, H. 2009. Differential approach to assess the validity of a motion-based driving simulator. *Journal of Virtual Reality and Broadcasting*, 6.

Nelder, J.A. and Mead, R. 1965. A simplex method for function minimization. *Computer Journal*, 7, 308–313.

Newell, G. 1993. A simplified theory of kinematic waves in highway traffic I: general theory. *Transportation Research Part B*, 27(4), 281–287.

Newell, G.F. 2002. A simplified car-following theory: a lower order model. *Transportation Research Part B*, 36(3), 195–205.

Ngoduy, D. and Hoogendoorn, S.P. 2003. An automated calibration procedure for macroscopic traffic flow models. Proceedings of 10th IFAC CTS, Tokyo.

NGSIM, 2011. Community Home of the Next Generation SIMulation Community. http://ngsim-community.org/

Nihan, N., Zhang, X., and Wang, Y. 2002. Dual Loop Error Evaluation of Dual Loop Data Accuracy Using Video Ground Truth Data. Research Report T1803 sponsored by Washington State Transportation, Northwest Transportation Commission, University of Washington, and U.S. Department of Transportation. March.

Oketch, T. and Carrick, M. 2005. Calibration and validation of a micro-simulation model in network analysis. Proceedings of 84rd Annual Meeting. Washington: Transportation Research Board.

Okutani, I. 1987. The Kalman filtering approaches in some transportation and traffic problems. Proceedings of 10th International Symposium on Transportation and Traffic Theory.

Ortuzar, J. and Willumsen, L.G. 2004. *Modeling Transport*, 3rd ed. New York: Wiley.

Ossen, S.J.L. and Hoogendoorn, S.P. 2005. Car-following behavior analysis from microscopic trajectory data. *Transportation Research Record*, 1934, 13–21.

Ossen, S.J.L. and Hoogendoorn, S.P. 2008a. Validity of trajectory-based calibration approach of car-following models in presence of measurement errors. *Transportation Research Record*, 2088, 117–125.

Ossen, S.J.L. and Hoogendoorn, S.P. 2008b. Calibrating car-following models using microscopic trajectory data: a critical analysis of both microscopic trajectory data collection methods, and calibration studies based on these data. Report. Delft University of Technology.

Ossen, S.J.L. and Hoogendoorn, S.P. 2009. Reliability of parameter values estimated using trajectory observations. *Transportation Research Record*, 2124, 36–44.

Ossen, S.J.L. 2008. Longitudinal driving behavior: theory and empirics. PhD thesis. Delft University of Technology.

Ossen, S.J.L., Hoogendoorn, S.P., and Gorte, B.G.H. 2006. Interdriver differences in car-following: a vehicle trajectory-based study. *Transportation Research Record*, 1965, 121–129.

Ou, Q. 2011. Fusing heterogeneous traffic data: parsimonious approaches using data–data consistency. PhD thesis. Delft University of Technology, Netherlands.

Papageorgiou, M. 1998. Some remarks on macroscopic traffic flow modeling. *Transportation Research Part A*, 32(5), 323–329.

Park, B., Pampati, D.M., and Balakrishna, R. 2006. Architecture for on-line deployment of DynaMIT in Hampton Roads, VA. In *Proceedings of 9th International Conference on Applications of Advanced Technology in Transportation*, pp. 605–610.

Park, B. and Qi, H. 2005. Development and evaluation of simulation model calibration procedure. *Transportation Research Record*, 1934, 208–217.

Park, B. and Schneeberger, J.D. 2003. Microscopic simulation model calibration and validation: case study of VISSIM simulation model for a coordinated signal system. *Transportation Research Record*, 1856, 185–192.

Patel, I., Kumar, A., and Manne, G. 2003. Sensitivity analysis of CAL3QHC Roadway Intersection Model. *Transportation Research Record*, 1842, 109–117.

Pipes, L.A. 1967. Car-following models and the fundamental diagram of road traffic, *Transportation Research*, 1, 21–29.

Plackett, R.L. 1950. Some theorems in least squares. *Biometrika*, 37(1–2), 149–157.

Powell, M.J.D. 1973. On search directions for minimization algorithms. *Mathematical Programming*, 4(1), 193–201.

Press, W.H., Teukolsky, S.A., Vetterling, W.T., et al. 1992. *Numerical Recipes in C*, 2nd ed. Cambridge, UK: Cambridge University Press.

PTV. 2006. VISUM 11.5. http://www.ptvag.com/software/transportation-planning-traffic-engineering/software-system-solutions/visum/

Punzo, V., Borzacchiello, M.T., and Ciuffo, B. 2011. On the assessment of vehicle trajectory data accuracy and application to next generation SIMulation (NGSIM) program data. Transportation Research Part C 19, 1243–1262.

Punzo, V. and Ciuffo, B. 2009. How parameters of microscopic traffic flow models relate to traffic conditions and implications on model calibration. *Transportation Research Record*, 2124, 249–256.

Punzo, V. and Ciuffo, B. 2010. Sensitivity analysis in car following models. Proceedings of Traffic Flow Theory Committee Summer Meeting and Conference, Annecy, France, July 7–9.

Punzo, V. and Ciuffo, B. 2011. Sensitivity analysis of microscopic traffic flow models: methodology and application. Proceedings of 90th Annual Meeting. Washington, DC: Transportation Research Board.

Punzo, V., Ciuffo, B., and Montanino, M. 2012. Can results of car-following model calibration based on trajectory data be trusted? Transportation Research Record, 2315, 11–24.

Punzo, V., Formisano, D.J., and Torrieri, V. 2005. Nonstationary Kalman filter for estimation of accurate and consistent car-following data. *Transportation Research Record*, 1934, 3–12.

Punzo, V. and Simonelli, F. 2005. Analysis and comparison of car-following models using real traffic microscopic data. *Transportation Research Record*, 1934, 53–63.

Punzo, V. and Tripodi, A. 2007. Steady-state solutions and multi-class calibration of Gipps' microscopic traffic flow model. *Transportation Research Record*, 1999, 104–114.

Qin, X. and Mahmassani, H. 2004. Adaptive calibration of dynamic speed density relations for online network traffic estimation and prediction applications. Proceedings of 83rd Annual Meeting. Washington, DC: Transportation Research Board.

Quadstone Ltd. 2003. *PARAMICS Version 4.2 Modeler Reference Manual*. Edinburgh.

Quendler, E., Kristler, I., Pohl, A., et al. 2006. Driver Assistance System LISA: Intelligent Infrastructure. Workshop of BMVIT 3 (in German).

Rakha, H., Hellinga, B., Van Aerde, M., et al. 1996. Systematic verification, validation and calibration of traffic simulation models. Proceedings of 75th Annual Meeting. Washington: Transportation Research Board, Washington, DC.

Ramsay, J.O. and Silverman, B.W. 1997. *Functional Data Analysis.* Heidelberg: Springer.

Ranjitkar, P., Nakatsuji, T., and Asanom M. 2004. Performance evaluation of microscopic flow models with test track data. *Transportation Research Record*, 1876, 90–100.

Rao, L. and Owen, L. 2000. Validation of high-fidelity traffic simulation models. *Transportation Research Record*, 1710, 69–78.

Rao, L., Goldsman, D., and Owen, L. 1998. Development and application of a validation framework for traffic simulation models. In *Proceedings of 1998 Winter Simulation Conference 2*, pp. 1079–1086.

Reason, J.T. and Brand, J.J. 1975. *Motion Sickness.* London: Academic Press.

Reimer, B., D'Ambrosio, L., Coughlin, J., et al. 2006. Using self-reported data to assess the validity of driving simulator data. *Behavior Research Methods*, 38(2), 314–324.

Remias, R., Hainen, A., Mitkey, S., et al. 2011. Probe Vehicle Re-Identification Data Accuracy Evaluation. Research Report. West Lafayette, IN: Purdue University.

Richards, P. 1956. Shock waves on the highway. *Operations Research*, 4(1), 42–51.

Riemersma, J., Van der Horst, R., Hoekstra, W., et al. 1990. The validity of a driving simulator in evaluating speed reducing measures. *Traffic Engineering and Control*, 31(7–8), 416–420.

Rilett, L.R., Kim, K., and Raney, B. 2000. A comparison of the low fidelity TRANSIMS and high fidelity CORSIM highway simulation models with intelligent transportation system data. *Transportation Research Record*, 1739, 1–8.

Ritchie, S.G. and Sun, C. 1998. Section-Related Measures of Traffic System Performance: Final Report. UCB-ITS-PRR-98-33, Sacramento: California PATH.

Rizzo, M., McGehee, D.V., Dawson, J.D., et al. 2001. Simulated car crashes at intersections in drivers with Alzheimer's disease. *Alzheimer's Disease and Associated Disorders*, 15(1), 10–20.

Saltelli, A. 2002. Making best use of model evaluations to compute sensitivity indices. *Computer Physics Communications*, 145(2), 280–297.

Saltelli, A., Annoni, P., Azzini, I., et al. 2010. Variance-based sensitivity analysis of model output: design and estimator for the total sensitivity index. *Computer Physics Communications*, 181(2), 259–270.

Saltelli, A. and Bolado, R. 1998. An alternative way to compute Fourier amplitude sensitivity tests. *Computational Statistics and Data Analysis*, 26(4), 445–460.

Saltelli, A., Ratto, M., Anres, T., et al. 2008. *Global Sensitivity Analysis: The Primer.* Chichester: Wiley.

Saltelli, A., Ratto, M., Tarantola, S., et al. 2006. Sensitivity analysis practices: strategies for model-based inference. *Reliability Engineering and System Safety*, 91(10–11), 1109–1125.

Saltelli, A. and Tarantola, S. 2002. On the relative importance of input factors in mathematical models: safety assessment for nuclear waste disposal. *Journal of American Statistical Association*, 97(459), 702–709.

Saltelli, A., Tarantola, S., Campolongo, F., et al. 2004. *Sensitivity Analysis in Practice: A Guide to Assessing Scientific Models*. Chichester: Wiley.

Saltelli, A., Tarantola, S., and Chan, K.P.S. 1999 A quantitative model-independent method for global sensitivity analysis of model output. *Technometrics*, 41(1), 39–56.

Scheffé, H. 1970. Practical solutions of the Behrens-Fisher problem. *Journal of American Statistical Association*, 65(332), 1501–1508.

Schlaich, J., Otterstätter, T., and Friedrich, M. 2010. Generating trajectories from mobile phone data. Proceedings of 89th Annual Meeting. Washington, DC: Transportation Research Board.

Schönhof, M. and Helbing, D. 2007. Empirical features of congested traffic states and their implications for traffic modeling. *Transportation Science*, 41(2), 135–166.

Schönhof, M. and Helbing, D. 2009. Criticism of three-phase traffic theory. *Transportation Research Part B*, 43(7), 784–797.

Schreiter, T. 2013, Vehicle-class Specific Control of Freeway Traffic, PhD thesis, Delft University of Technology.

Schreiter, T., Van Lint, J.W.C., Yuan, Y. and Hoogendoorn, S.P. et al. 2010. Propagation wave speed estimation of freeway traffic with image processing tools. Proceedings of 89th Annual Meeting. Washington, DC: Transportation Research Board.

Schruben, L. 1980. Confidence interval estimation using standardized time series. *Operations Research*, 31(6), 1090–1108.

Schultz, G.G. and Rilett, L.R. 2004. Analysis of distribution and calibration of car-following sensitivity parameters in microscopic traffic simulation models. *Transportation Research Record*, 1876, 41–51.

Schultz, G.G. and Rilett, L.R. 2005. Calibration of distributions of commercial motor vehicles in CORSIM. *Transportation Research Record*, 1934, 246–255.

SENSOR 2004. *Secondary Road Network Traffic Management Strategies: Handbook for Data Collection, Communication and Organisation*. Document D1.8d. Brussels: European Commission.

Shaaban, K.S. and Radwan, E. 2005. A calibration and validation procedure for microscopic simulation model: a case study of SimTraffic for arterial streets. Proceedings of 84th Annual Meeting. Washington, DC: Transportation Research Board.

Shaffer, J.P. 1995. Multiple hypothesis testing. *Annual Review of Psychology*, 46, 561–584.

Shapiro, V., Gluhchev, G., and Dimov, D. 2006. Toward a multinational car license plate recognition system. *Machine Vision and Applications*, 17(3), 173–183.

Sheffi, Y. 1985. *Urban Transportation Networks*. Englewood Cliffs, NJ: Prentice Hall.

Sherali, H.D. and Park, T. 2001. Estimation of dynamic origin–destination trip tables for a general network. *Transportation Research Part B*, 35(3), 217–235.

SIMLAB. 2010. http://simlab.jrc.ec.europa.eu/

Simonelli, F., Papola, A., Marzano, V. vitiello, I. 2011. A methodology for locating link count sensors taking into account the reliability of prior OD matrix estimates. *Transportation Research Record* Nr 2263, pp 182–190, 2011.

Smith, M., Duncan, G., and Druitt, S. 1995. PARAMICS: microscopic traffic simulation for congestion management. *In IEEE Colloquium on Dynamic Control of Strategic Inter-Urban Road Networks*, pp. 8/1–8/3.

Sobol, I.M. 1976. Uniformly distributed sequences with an additional uniform property. *Computational Mathematics and Mathematical Physics*, 16(5), 236–242.

Sobol, I.M. 1993. Sensitivity analysis for non-linear mathematical models. *Mathematical Models and Computational Experiments*, 1, 407–414.

Sobol, I.M. 2001. Global sensitivity indices for non-linear mathematical models and their Monte Carlo estimates. *Mathematics and Computers in Simulation*, 55(1-3), 271–280.

Sobol, I.M., Tarantola, S., Gatelli, D. et al., 2007. Estimating the approximation error when fixing unessential factors in global sensitivity analysis. *Reliability Engineering and System Safety*, 92(7), 957–960.

Soriguera, F. and Robusté, F. 2011. Estimation of traffic stream space mean speed from time aggregations of double loop detector data. *Transportation Research Part C*, 19(1), 115–129.

Soriguera, F. and Robusté, F. 2011. Highway travel time accurate measurement and short-term prediction using multiple data sources. *Transportmetrica*, 7(1), 85–109.

Spall, J.C. 1992. Multivariate stochastic approximation using a simultaneous perturbation gradient approximation. *IEEE Transactions on Automatic Control*, 37(3), 332–341.

Spall, J.C. 1998. Implementation of the simultaneous perturbation algorithm for stochastic optimization. *IEEE Transactions on Aerospace and Electronic Systems*, 34(3), 817–823.

Spall, J.C. 2000. Adaptive stochastic approximation by the simultaneous perturbation method. *IEEE Transactions on Automatic Control*, 45(10), 1839–1853.

Spall, J.C. 2001. *A Method for System Optimization*. Baltimore: Johns Hopkins University, www.jhuapl.edu/SPSA/

Spall, J.C. 2003. *Introduction to Stochastic Search and Optimization. Estimation, Simulation and Control*. Hoboken, NJ: Wiley.

Spall, J.C. 2009. Feedback and weighting mechanisms for improving Jacobian estimates in the adaptive simultaneous perturbation algorithm. *IEEE Transactions on Automatic Control*, 54(6), 1216–1229.

Spall, J.C., Hill, S., and Stark, D.R. 2006. Theoretical framework for comparing several stochastic optimization approaches. In G. Calafiore and F. Dabbene, Eds., *Probabilistic and Randomized Methods for Design under Uncertainty*. Heidelberg: Springer, pp. 99–117.

Spiess, H. 1987. A maximum likelihood model for estimating origin–destination matrices. *Transportation Research Part B*, 21(5), 395–412.

Spiess, H. 1990. A gradient approach for the OD matrix adjustment problem. Publication 693. Transportation Research Centre, Montreal University.

Stipdonk, H.L. 2013. Road safety in bits and pieces; for a better understanding of the development of the number of road fatalities. PhD thesis, Delft University of Technology.

Stipdonk, H., Van Toorenburg, J., and Postema, M. 2008. Phase diagram distortion from traffic parameter averaging. Proceedings of European Transport Conference.

Sun, C., Arr, G., Ramachandran, R., et al. 2004. Vehicle re-identification using multidetector fusion. *IEEE Transactions on Intelligent Transportation Systems*, 5(3), 155–164.

Sundaram, S. 2002. Development of a dynamic traffic assignment system for short-term planning applications. Master's thesis. Cambridge: Massachusetts Institute of Technology.

Szeto, M.W. and Gazis, D.C. 1972. Application of Kalman filtering to the surveillance and control of traffic systems. *Transportation Science*, 6(4), 419–439.

Tabib, S. 2001. Vehicle re-identification based on inductance signature matching. Master's thesis. University of Toronto.

Tavana, H. 2001. Internally consistent estimation of dynamic network origin–destination flows from intelligent transportation systems data using bilevel optimization. PhD thesis. Austin: University of Texas.

Tavana, H. and Mahmassani, H. 2000. Estimation of dynamic origin–destination flows from sensor data using bilevel optimization. Proceedings of 79th Annual Meeting. Washington, DC: Transportation Research Board.

Theil, H. 1961. *Economic Forecasts and Policy*. Amsterdam: North-Holland.

Thiemann, C., Treiber, M., and Kesting, A. 2008. Estimating acceleration and lane-changing dynamics based on NGSIM trajectory data. *Transportation Research Record*, 2088, 90–101.

Tobin, R. and Friesz, T. 1988. Sensitivity analysis for equilibrium network flow. *Transportation Science*, 22(4), 242–250.

Toledo, T., Ben-Akiva, M.E., Darda, D., et al. 2004. Calibration of microscopic traffic simulation models with aggregate data. *Transportation Research Record*, 1876, 10–19.

Toledo, T. and Farah, H. 2011. Alternative definitions of passing critical gaps. *Transportation Research Record*, 2260, 76–82.

Toledo, T. and Koutsopoulos, H.N. 2004. Statistical validation of traffic simulation model. *Transportation Research Record*, 1876, 142–150.

Toledo, T., Koutsopoulos, H.N., and Ahmed, K.I. 2007. Estimation of vehicle trajectories with locally weighted regression. *Transportation Research Record*, 1999, 161–169.

Toledo, T., Koutsopoulos, H.N., Davol, A., et al. 2003. Calibration and validation of microscopic traffic simulation tools: Stockholm case study. *Transportation Research Record*, 1831, 65–75.

TomTom, 2011. Traffic Stats, http://trafficstats.tomtom.com//

Tornos, J. 1998. Driving behaviour in a real and a simulated road-tunnel: a validation study. *Accident Analysis and Prevention*, 30(4), 497-503.

Treiber, M. and Helbing, D. 2002. Reconstructing spatio-temporal traffic dynamics from stationary detector data. *Cooperative Transportation Dynamics*, 1(3), 3.1–3.21.

Treiber, M., Hennecke, A., and Helbing, D. 2000. Congested traffic states in empirical observations and microscopic simulations. *Physical Reviews E*, 62(2), 1805–1824.

Treiber, M. and Kesting, A. 2012. Validation of traffic flow models with respect to the spatiotemporal evolution of congested traffic patterns. *Transportation Research Part C*, 21(1), 31–41.

Treiber, M., Kesting, A., and Helbing, D. 2006. Delays, inaccuracies, and anticipation in microscopic traffic models. *Physica A*, 360(1), 71–88.

Treiber, M., Kesting, A., and Wilson, R.E. 2011. Reconstructing the traffic state by fusion of heterogeneous data. *Computer-Aided Civil and Infrastructure Engineering*, 26, 408–419.

Transport Simulation Systems. 2008. *AIMSUN 6 Microsimulator User's Manual*.

Turing, A.M. 1950. Computing machinery and intelligence, *Mind*, 59, 433–460.

Ugray, Z., Lasdon, L., Plummer, J., et al. 2005. A multistart scatter search heuristic for smooth NLP and MINLP problems. In R. Sjarda et al., Eds., *Metaheuristic Optimization via Memory and Evolution*. Heidelberg: Springer, pp. 25–57.

Valero-Mora, P.M., Tontsch, A., Welsh, R., et al. 2013. Is naturalistic driving research possible with highly instrumented cars? Lessons learnt in three research centres. *Accident Analysis and Prevention*, 58, 187–194.

Van Aerde, M. and Rakha, H. 1995. TRAVTEK Evaluation Modeling Study. Technical Report. U.S. Department of Transportation, Federal Highway Administration.

Van Arem, B. and Van der Vlist, M.J.M. 1992. An On–Line Procedure for Estimating Current Capacity. Technical Report INRO-VVG 1991-17, TNO Institute of Spatial Organization, Delft.

Van Lint, J. 2010. Empirical evaluation of new robust travel time estimation algorithms. *Transportation Research Record*, 2160, 50–59.

Van Lint, J.W.C. 2003. Confidence intervals for real-time freeway travel time prediction. *Proceedings of IEEE Intelligent Transportation Systems*, 2, 1453–1458.

Van Lint, J.W.C. 2004. Reliable travel time prediction for freeways. PhD thesis. Delft University of Technology.

Van Lint, J.W.C. and Hoogendoorn, S.P. 2007. The technical and economic benefits of data fusion for real-time monitoring of freeway traffic. World Congress of Transportation Research, Berkeley, CA.

Van Lint, J.W.C. and Hoogendoorn, S.P. 2009. A robust and efficient method for fusing heterogeneous data from traffic sensors on freeways. *Computer Aided Civil and Infrastructure Engineering*, 25(8), 596–612.

Van Lint, J.W.C., Schreiter, T., and Yuan, Y. 2009. Technische en functionele haalbaarheid check-algoritme voor de productie van statistische verkeersgegevens en indicatoren. Technical University of Delft.

Van Lint, J.W.C. and Van der Zijpp, N.J. 2003. Improving a travel time estimation algorithm by using dual loop detectors. *Transportation Research Record*, 1855, 41–48.

Van Vliet, D. 2009. *SATURN: Simulation and Assignment of Traffic in Urban Road Networks. Manual*, Version 10.9. Leeds. UK: ITS.

Van Winsum, W. 1996. Speed choice and steering behavior in curve driving. *Human Factors*, 38(3), 434–441.

Van Winsum, W. and Brouwer, W. 1997. Time headway in car following and operational performance during unexpected braking. *Perceptual and Motor Skills*, 84(3), 1247–1257.

Van Zuylen, H.J. 1979. The estimation of turning flows on a junction. *Traffic Engineering and Control*, 20(11), 539–541.

Van Zuylen, H.J. and Willumsen, L.G. 1980. The most likely trip estimated from traffic counts. *Transportation Research Part B*, 14(3), 281–293.

Vandekerckhove, J. 2010. General simulated annealing algorithm. http://www. mathworks.it/matlabcentral/fileexchange/10548-general-simulated-annealing-algorithm

Vardi, Y. 1996. Network tomography: estimating source-destination traffic intensities from link data. *Journal of American Statistical Association*, 91(433), 365–377.

Varshney, P.K. 1997. Multisensor data fusion. *Electronics and Communications Engineering Journal*, 9(6), 245–253.

Vaze, V., Antoniou, C., Wen, Y. et al. 2009. Calibration of dynamic traffic assignment models with point-to-point traffic surveillance. *Transportation Research Record*, 2090, 1–9.

Versavel, J. 2007. Traffic data collection: Quality aspects of video detection. Proceedings of 86th Annual Meeting. Washington, : Transportation Research Board.

Wang, Y. and Papageorgiou, M. 2005. Real-time freeway traffic state estimation based on extended Kalman filter: a general approach. *Transportation Research Part B*, 39(2), 141–167.

Wang, Y., Papageorgiou, M. and Messmer, A. 2006. A real-time freeway network traffic surveillance tool. *IEEE Transactions on Control Systems Technology*, 14(1), 18–32.

Wang, Y., Papageorgiou, M., and Messmer, A. 2007. Real-time freeway traffic state estimation based on extended Kalman filter: a case study. *Transportation Science*, 41(2), 167–181.

Wang, Y., Papageorgiou, M., and Messmer, A. 2008. Real-time freeway traffic state estimation based on extended Kalman filter: adaptive capabilities and real data testing. *Transportation Research Part A*, 42(10), 1340–1358.

Wang, Y., Papageorgiou, M., Messmer, A. et al. 2009. An adaptive freeway traffic state estimator. *Automatica*, 45(1), 10–24.

Wikipedia. 2011. Electronic toll collection. http://en.wikipedia.org/wiki/Electronic_toll_collection

Willsky, A.S., Chow, E.Y., Gershwin, S.B., et al. 1980. Dynamic model-based techniques for the detection of incidents on freeways. *IEEE Transactions on Automatic Control*, 25(3), 347–360.

Wilson, R.E. 2001. An analysis of the Gipps' car-following model of highway traffic. *IMA Journal of Applied Mathematics*, 66(5), 509–537.

Windover, J. and Cassidy, M.J. 2001. Some observed details of freeway traffic evolution. *Transportation Research Part A*, 35(10), 881–894.

Wu, N. 2002. A new fundamental approach for modeling of fundamental diagrams. *Transportation Research Part A*, 36(10), 867–884.

Xin, W., Hourdos, J., and Michalopoulos, P. 2008. A vehicle trajectory collection and processing methodology and its implementation to crash data. Proceedings of 87th Annual Meeting. Washington: Transportation Research Board.

Yan, X., Abdel-Aty, M., Radwan, E., et al. 2008. Validating a driving simulator using surrogate safety measures. *Accident Analysis and Prevention*, 40(1), 274–288.

Yang, H. 1995. Heuristic algorithms for the bilevel origin–destination matrix estimation algorithm. *Transportation Research Part B*, 29(4), 231–242.

Yang, H. 1997. Sensitivity analysis for the elastic demand network equilibrium problem with application. *Transportation Research Part B*, 31(1), 55–70.

Yang, H. 1998. Sensitivity analysis for queuing equilibrium network flow and its application to traffic control. *Mathematical and Computer Modeling*, 22(4–7), 247–258.

Yang, H., Gan, L., and Tang, W.H. 2001. Determining cordons and screen lines for origin-destination trip studies. *Proceedings of Eastern Asia Society for Transportation Studies*, 3(2), 85–99.

Yang, H., Iida, Y., and Sasaki, T. 1994. The equilibrium-based origin–destination matrix estimation problem. *Transportation Research Part B*, 28(1), 23–33.

Yang, H., Meng, Q., and Bell, M.G.H. 2001a. Simultaneous estimation of the origin–destination matrices and travel-cost coefficient for congested networks in a stochastic user equilibrium. *Transportation Science*, 35(2), 107–123.

Yang, H., Sasaki, T., Iida, Y., et al. 1992. Estimation of origin–destination matrices from traffic counts on congested networks. *Transportation Research Part B*, 26(6), 417–434.

Yang, H. and Yagar, S. 1995. Traffic assignment and signal control in saturated road networks. *Transportation Research Part A*, 29(2), 125–139.

Yang, H., Yang, C., and Gan, L. 2006. Models and algorithms for the screen line-based traffic counting locations problems. *Computers and Operations Research*, 33(3), 836–858.

Yang, H. and Zhou, J. 1998. Optimal traffic counting locations for origin–destination matrix estimation. *Transportation Research Part B*, 32(2), 109–126.

Yang, Q. and Koutsopoulos, H.N. 1996. A microscopic traffic simulator for evaluation of dynamic traffic management systems. *Transportation Research Part C*, 4(3), 113–129.

Young, F.W. and Hamer, R.M. 1987. *Multidimensional Scaling: History, Theory, and Applications*. Mahwah, NJ: Lawrence Erlbaum.

Young, P.C., Parkinson, S., and Lees, M. 1996. Simplicity out of complexity in environmental modeling: Occam's razor revisited. *Journal of Applied Statistics*, 23(2–3), 165–210.

Yu, X. 2009. Evaluation of non-intrusive sensors for vehicle classification on freeways. Student Competition, 2nd International Symposium on Freeway and Tollway Operations. Hawaii.

Yuan, Y. 2012. Lagrangian multiclass traffic state estimation. PhD thesis. Delft University of Technology.

Yue, P. and Yu, L. 2000. Travel Demand Forecasting Models: a Comparison of EMME/2 and QRS II Using a Real-World Network. Technical Report. Houston: Texas Southern University Center for Transportation Training and Research.

Zhang, L., Kovvali, V., Clark, N. et al. 2007. *NGSIM-VIDEO User's Manual*. Publication FHWA-HOP-07-009. U.S. Department of Transportation.

Zhang, X. and Rice, J. 1999. Visualizing Loop Detector Data. Technical Report. Berkeley: University of California.

Zheng, Z., Ahn, S., Chen, D. et al. 2011a. Applications of wavelet transform for analysis of freeway traffic: Bottlenecks, transient traffic, and traffic oscillations. *Transportation Research Part B*, 45(2), 372–384.

Zheng, Z., Ahn, S., Chen, D. et al. 2011b. Freeway traffic oscillations: microscopic analysis of formations and propagations using wavelet transform. *Transportation Research Part B*, 45(2), 372–384.

Zhou, X., Qin, X., and Mahmassani, H.S. 2003. Dynamic origin–destination demand estimation using multi-day link traffic counts for planning applications. *Transportation Research Record*, 1831, 30–38.

Appendix A: Glossary

Assignment   Selection of routes between origins and destinations in a transport network.

Calibration   Procedure whereby appropriate values of model parameters are estimated, in other words, model fitting or parameter estimation.

Car following model   Model of vehicle response in a flow of traffic based on behavior of its predecessor.

Data collection   Process of gathering required data to prepare inputs for running a simulation model.

Data enhancement   (1) Generation of new variables or data sets from those initially collected. (2) Extension of existing data set by replacing missing values or correcting wrong ones.

Data fusion   Linkage of different elements of data in a coherent manner.

Data processing   Specifying and fitting (formal statistical) models to estimate parameters, test hypotheses, predict traffic states, and visualize outputs.

Density   Number of vehicles or units of traffic per unit length of road network.

Detector   Device that collects traffic information; may utilize different technologies and record data with varying degrees of fidelity, accuracy, and reliability; may be fixed at location or move with a vehicle.

Deterministic   Nonstochastic; evolution of a system is predictable and repeatable.

Dynamic   Evolving over time; not static.

Error (systematic and stochastic)   Difference between estimate and true value; a systematic is biased (consistently positive or negative); a stochastic error changes over time without a predetermined pattern.

Filtering   Process of treating measured data so that the values become meaningful and lie within a range of expected values.

Flow   Number of vehicles or units of traffic passing pass a cross section in a unit of time.

Fundamental diagram   Representation of flow versus density, speed versus flow, or speed versus density. The two traffic states distinguishable in the flow–density relationship are free flow and congestion.

Gap acceptance model   Model of driver behavior during crossing or merging to a traffic flow and achieving a lane change.

Goodness of fit   Measure of distance between observed and modeled values of a given measure of performance.

Jam density   Maximum density of a network element.

Lane changing model   Model of behavior of a driver who wants to change lanes. This process has two phases: (1) making the decision whether to change lanes and (2) the lane changing operation.

Macroscopic   Model using aggregate description of a modeled phenomenon.

Measure of performance   Variable chosen to assess performance of a modeled situation. This choice is to be made in agreement with the aim of the policy or regulation in question.

Mesoscopic   Model using a combination of aggregate and disaggregate descriptions of a modeled phenomenon.

Microscopic   Model using disaggregate description of a modeled phenomenon.

Network coding   Development of necessary inputs for network configuration as a preparatory step for simulation.

OD matrix   Matrix of trips between pairs of zones of origin (O) and destination (D).

Parameter   Element of a model that can be used to tweak its performance or ability to reproduce observed reality; computation from data values recorded but not actually a data value recorded from a subject.

Sensitivity analysis   Process for determining how different values of a variable will impact other variable(s) under a set of alternative reasonable assumptions.

Shock wave speed   Speed at which a shock wave propagates.

Simulation   Process by which a model is run to produce a representation of an actual or predicted situation.

Smoothing   Process of capturing important patterns of a dataset by damping the noise.

Space mean speed   Arithmetic mean of speeds of vehicles present on a road section at a given moment.

Static   Not evolving over time (opposite of dynamic).

Stochastic   Nondeterministic; its subsequent state is determined by both the predictable actions of the process and a random element.

Time mean speed   Arithmetic mean of speeds of vehicles passing a cross section along a roadway measured over a fixed time period.

Trajectory   Path followed by a vehicle over time, usually represented in the (x,t) plane.

Validation   Verification that model behavior accurately represents the real world system modeled.

Variable   Symbolic name assigned to a value whose associated value may be changed.

Appendix B: Application of variance-based methods to microscopic traffic flow models

CONTENTS

In this appendix, a detailed illustration of a sensitivity analysis study is provided. In particular, variance-based techniques are applied to two well known car-following models in a *factor fixing setting*. The objective of the study is to identify the less influential parameters that may be fixed at any reasonable value without significantly affecting the model results or, conversely, the subset of parameters to which limit the model calibration. Due to the complexity and the cost of any (simulation) model calibration, this result is deemed to be crucial in the practical applications of (traffic) models.

All the study steps are described following the general framework for uncertainty management described at the beginning of the Chapter 5 and depicted in Figure 5.1. Most of the material is taken from Ciuffo et al. (2014).

B.1 STEP A: PROBLEM SPECIFICATION

B1.1 Model description

The analysis is applied to two well-known car-following models: the intelligent driver model (IDM; Treiber et al., 2000) and the Gipps model (Gipps, 1981). We use the same notations for both models to describe the variables and the parameters with similar meanings. In particular, the Greek letters α, β, and γ indicate the model parameters that are generally considered fixed in the field literature and whose same constant values are usually adopted in simulation practice. Apart from these parameters, the following common notations are adopted:

> Subscripts f and l indicate the follower and the leader, respectively.
> $s(t)$, $\dot{s}(t)$, and $\ddot{s}(t)$ represent the state variables: vehicle's is position, speed, and acceleration, respectively, as functions of time t.
> $\Delta s(t) = s_l(t) - s_f(t)$ indicates the space between the leader's front bumper and the follower's front bumper at time t.
> $\dot{S}$ and $\ddot{S}$ denote maximum speed and maximum acceleration (parameters).
> L_l indicates the leader's vehicle length (parameter).
>
> ΔS^0 represents the rear end–front desired inter-vehicle spacing at stop (parameter).

Finally, for the sake of simplicity the common notation b_f is adopted for the parameter representing the follower's deceleration. However, this notation has different meanings in the two models. In the IDM it represents a "comfortable" deceleration. In the Gipps model, it indicates the "most severe braking that the driver of the following vehicle wishes to undertake" (Gipps, 1981).

B1.1.1 Intelligent driver model (IDM)

The IDM belongs to the class of the social force models. It is developed by an analogy of the molecular dynamics method (Helbing and Tilch, 1998). The social force concept states that driving behavior is motivated by a sum of social forces, including the force that pushes the vehicle to reach the driver's desired speed, and the interaction force that compels the vehicle to keep a suitable distance from the previous vehicle (Wang et al., 2010). The IDM is a simple car-following model with descriptive parameters (Treiber et al., 2006) that returns the acceleration of a driver as a continuous function of the vehicle's speed, distance, and relative speed with respect to the preceding vehicle. The model formulation is the following:

$$\ddot{s}(t) = \ddot{S}_f \cdot \left[1 - \left(\frac{\dot{s}_f(t)}{\dot{S}_f} \right)^\alpha - \left(\frac{\Delta S^*(t)}{\Delta s(t) - L_l} \right)^\beta \right] \tag{B.1}$$

with:

$$\Delta S^*(t) = \Delta S^0 + \max\left(T \cdot \dot{s}_f(t) + \frac{\dot{s}_f(t) \cdot (\dot{s}_f(t) - \dot{s}_l(t))}{2 \cdot \sqrt{\ddot{S}_f \cdot |b_f|}}, 0 \right) \tag{B.2}$$

where $\Delta S^*(t)$ is the desired distance from the leader and T is the minimum time headway. From the model observation, one can make the following remarks:

The parameter $\ddot{S}_f$ corresponds to the acceleration applied by the follower at start when the distance from the leader is much larger than the distance ΔS^0.

$\ddot{S}_f$ also corresponds to the deceleration of a vehicle traveling at its desired speed and whose distance from the leader approximates the desired distance.

$\dot{S}_f$ is the maximum desired speed [higher values are not allowed by Equation (B.2)] that provides a deceleration as soon as this value is exceeded).

Treiber et al. (2000) show that bf is a sort of threshold between normal and emergency braking conditions.

For what concerns the desired spacing ΔS^*, from Equation (B.2) we can derive the following:

When a vehicle is stopped (i.e., $\dot{s}_f(t) = 0$), the follower's desired spacing is equal to ΔS^0, which therefore has the meaning of a desired distance at stop.

At equilibrium (i.e., $\dot{s}_f(t) = \dot{s}_l(t)$), the follower's desired spacing is equal to ΔS^0 plus the space traveled by the follower in the minimum time headway, T.

In all other cases, the follower increases his or her desired spacing at equilibrium by an additional term for which it is possible to provide a physical interpretation: the difference of the leader's and the follower's stopping distances assuming an equal braking time and a follower's deceleration rate equal to $\sqrt{\ddot{S}_f \cdot |b_f|}$ (see Ciuffo et al. 2014, for an explanation). The max operator in Equation (B.2) is necessary for negative speed differences, i.e., $\dot{s}_f(t) < \dot{s}_l(t)$, to prevent the follower's desired distance from becoming lower than ΔS^0.

BI.I.2 Gipps car-following model

The Gipps car-following model is the most famous among the class of "safety distance" or "collision avoidance" models. The intent of models in this class is to specify a safe following distance and adapt driver's behavior to maintain the safe distance. In practice, the Gipps model assumes that the following driver chooses a speed to be able to maintain the minimum distance at standstill whenever the leader brakes at its maximum deceleration rate [Equation (B.5)]. In case the driver has no vehicles in front, the model defines a speed according to a free-flow acceleration profile to make the vehicle reach its maximum speed [Equation (B.4)].

Then, according to the Gipps model (1981), the speed planned by the follower at time t for the time $t + \tau$ (in which τ is the driver's apparent reaction time) is given by:

$$\dot{s}_f(t+\tau) = \max\{\min\{\dot{s}_{f,a}(t+\tau), \dot{s}_{f,b}(t+\tau)\}, 0\} \tag{B.3}$$

where:

$$\dot{s}_{f,a}(t+\tau) = \dot{s}_f(t) + \alpha \cdot \ddot{S}_f \cdot \tau \cdot \left(1 - \frac{\dot{s}_f(t)}{\hat{S}_f}\right) \cdot \left(\beta + \frac{\dot{s}_f(t)}{\hat{S}_f}\right)^\gamma \tag{B.4}$$

$$\dot{s}_{f,b}(t+\tau) = b_f \cdot \left(\frac{\tau}{2} + \theta\right)$$
$$+ \sqrt{b_f^2 \cdot \left(\frac{\tau}{2} + \theta\right)^2 - b_f \cdot \left[2 \cdot (\Delta s(t) - (L_l + \Delta S^0)) - \dot{s}_f(t) \cdot \tau - \frac{\dot{s}_f^2(t)}{\hat{b}}\right]} \tag{B.5}$$

in which $\hat{b}$ is the follower's estimate of the leader's most severe braking capabilities (deceleration parameters $\hat{b}$ and b_f are considered with their sign) and θ is an additional reaction time that prevents the follower from decelerating always at his or her maximum rate.

B.1.2 Uncertain inputs and model implementation

BI.2.I Intelligent driver model (IDM)

As anticipated in the previous subsection, parameters α and β in Equation (B.1) are generally considered as *fixed* in the field literature although only a value for β ($\beta = 2$) was given in the original paper by Treiber et al. (2000). Indeed, such value was indicated for the "intelligent" braking strategy to be effective, whereas the intended strategy for intelligent braking was the ability to resolve critical driving situations by dynamically reverting to a "safe" driving.

On the other hand, the authors demonstrated it was necessary for $\beta > 1$ to preserve the model from collision. According to the main objective of the application presented, all the model parameters, including α and β have been considered as *uncertain*, with the $\beta > 1$ condition imposed.

B1.2.2 Gipps car-following model

Equation (B.5) is a delayed differential equation that Gipps proposed to solve simply by adopting an integration step equal to the delay τ. Accordingly, a forward Euler method on acceleration, i.e., a trapezoidal integration scheme on speed has been here adopted for numerical resolution, with an integration step equal to τ.

Wilson (2001) thoroughly analyzed the stability of the Gipps car-following model. Wilson also showed that within the region of the input parameter space in which the model is unstable, its solution can lose global existence due to the occurrence of negative values under the square root in Equation (B.5). Wilson then conjectured that this condition arises if the equilibrium speed headway function of the follower is double valued as occurs when:

$$\dot{S}_f \geq \frac{\tau + \theta}{\frac{1}{|\hat{b}|} - \frac{1}{|b_f|}} \tag{B.6}$$

Having successfully verified such conjecture in a very high number of model runs in this study, only values of parameters that did not respect inequality (B.6) were preserved. It has to be said that this potentially introduces correlation among parameters that are the inputs of our sensitivity analysis. As anticipated, this condition imposes careful consideration in setting the analysis, as noted in the presentation of results for the Gipps model.

Within the intent of the study, also in the case of the Gipps model, all parameters were considered as *uncertain*, including α, β, and γ in Equation (B.4), which instead, in the original formulation and in the ensuing literature, were always considered as fixed and equal to 2.5, 0.025, and 0.5, respectively. In particular, such value combination guarantees that, independently from the value assumed by $\dot{s}_f(t)/\dot{S}_f$, the product of all the factors but $(\ddot{S}_f \cdot \tau)$ in the second term on the right side of Equation (B.4), never exceeds the unit value—preserving for $\ddot{S}_f$ the meaning of the maximum acceleration rate. Such values also implicitly define the specific value of the ratio $\dot{s}_f(t)/\dot{S}_f$ at which the maximum acceleration is reached, which means defining the vehicle engine's power curve. It is therefore clear that by considering such parameters uncertain (free to vary), the descriptive power of a model increases, making it possible to describe an engine's power curve.

This is a motivation in favor of the calibration of such parameters and provides a further reason to investigate their impacts on the outputs.[1]

All the considerations concerning the meanings of the parameters for the two car-following models are intended to drive the successive phases in which the variance-based sensitivity analysis techniques have been applied.

B.1.3 Fixed inputs

The fixed inputs are the leader trajectory and the initial position and speed of the follower. As such inputs are assumed certain as fixed and not left to vary, the following analysis will not inform the modeler on their significance in influencing the outputs. For this reason, the analysis was repeated twice, for two qualitatively different trajectories as shown in Figure B.1.

Such trajectories were obtained from a series of experiments carried out along roads in the area of Naples, Italy, under real traffic conditions between October 2002 and July 2003. Experiments were performed by driving four vehicles in a platoon along urban and extra-urban roads under different traffic conditions. All vehicles were equipped with kinematic GPS receivers that recorded their positions at 0.1-second intervals. Postprocessing of data involved the differential correction of raw GPS coordinates by means of the data gathered by a fifth stationary receiver acting as a base station and by the application of an elaborate filtering procedure. More details on the experimental set-up and the gathering of data can be found in Punzo and Simonelli (2005). The data filtering procedure is described in Punzo et al. (2005). The data are available to forum subscribers from the MULTITUDE Project Internet site (COST Action TU0903, 2011).

The trajectory data used herein are designated 30C and 30B and shown, respectively, at the upper left and right sides of Figure B.1. Set 30C consists of 6 minutes of data from a one-lane urban road. Set 30B is 4.2 minutes long and was taken from a two-lane rural highway.

---

[1] However, it is not necessary to relax all three parameters. By imposing the condition that vehicle acceleration never exceeds $\breve{S}$, it is possible to obtain the following relationship that replaces Equation (B.5):

$$\dot{s}_f(t+\tau) = \dot{s}_f(t) + \ddot{S}_f \cdot \tau \cdot \left(1 - \frac{\dot{s}_f(t)}{\dot{S}_f}\right) \cdot \left(\beta(\gamma) + \frac{\dot{s}_f(t)}{\dot{S}_f}\right)^{\gamma} \tag{B.7}$$

in which $\alpha = 1$ and β is replaced by a function of γ, $\beta(\gamma)$, which satisfies the condition imposed on vehicle acceleration. More details on Equation (B.7) and its derivation can be found in Ciuffo et al. (2012). It is also shown there that simply by making parameter γ uncertain, Equation (B.7) allows different speed–acceleration behaviors to be reproduced.

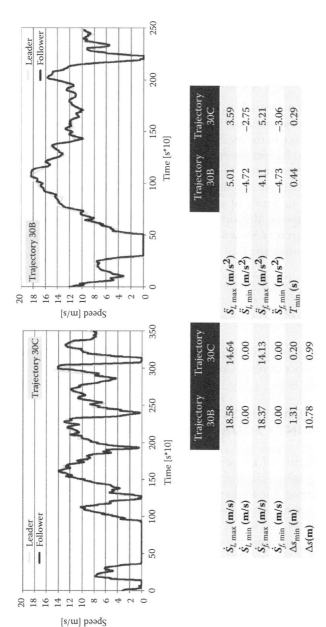

	Trajectory 30B	Trajectory 30C
$\dot{S}_{l,max}$ (m/s)	18.58	14.64
$\dot{S}_{l,min}$ (m/s)	0.00	0.00
$\dot{S}_{f,max}$ (m/s)	18.37	14.13
$\dot{S}_{f,min}$ (m/s)	0.00	0.00
Δs_{min} (m)	1.31	0.20
Δs(m)	10.78	0.99

	Trajectory 30B	Trajectory 30C
$\ddot{S}_{l,max}$ (m/s^2)	5.01	3.59
$\ddot{S}_{l,min}$ (m/s^2)	-4.72	-2.75
$\ddot{S}_{f,max}$ (m/s^2)	4.11	5.21
$\ddot{S}_{f,min}$ (m/s^2)	-4.73	-3.06
T_{min} (s)	0.44	0.29

Figure B.I Trajectories used in the analysis and main kinematic characteristics. Diagram on the left shows congested urban conditions (30C); extra-urban conditions are shown on the right. (30B). For the two trajectory pairs, maximum and minimum speed and acceleration, minimum time headway, minimum spacing, and spacing at the beginning of the simulation are reported. (*Source:* Ciuffo et al. 2013. With permission.)

From the available data, this study used only the trajectories of the first two vehicles of the platoon (see Figure B.1). As already noted, inputs to each car-following model consisted of the whole trajectory of the platoon leader and the initial position and speed of the first follower. The model output was therefore the whole trajectory of the follower.

B.1.4 Variables and quantity of interest

Concerning the output variables, two different measures were considered for the analysis: (1) vehicle speed and (2) inter-vehicle spacing of the leader. It is well known that the choice of the measure or quantity of interest affects the performances of the optimization algorithms and the results of calibration (which in our framework corresponds to Step B, i.e., the quantification of the uncertainties in the inputs). See Punzo et al. (2012a) for a discussion on this topic.

Dealing with time series of such measurements, the root mean square error ($RMSE$) between the simulated and actual follower's inter-vehicle spacing/speed (trajectory shown in Figure B.1) was adopted as the quantity of interest for the analysis:

$$RMSE(x,y) = \sqrt{\frac{1}{H}\sum_{i=1}^{H}(x_i - y_i)^2} \qquad (B.8)$$

where x and y are, respectively, the simulated and observed measurements and H is the number of observations in the trajectory.

B.2 STEP B: QUANTIFICATION OF UNCERTAINTY SOURCES IN INPUTS

Because the objective of the sensitivity analysis in this study is to understand how uncertainties in the outputs are apportioned to the inputs, the phase of quantifying the uncertainties in the inputs is crucial. In general, it is also the most subjective and difficult phase of the entire uncertainty management process. As noted earlier, in a probabilistic setting this phase generally implies defining the joint probability density functions (pdfs) of the uncertain inputs or their marginal pdfs with simplified correlation structures or even independence assumptions. It involves gathering information via direct observations, expert judgments, physical means, or indirect estimation.

In this study, we sought to make the assumed marginal pdfs of the model parameters independent. It is worth noting that although the assumption of parameter independence is likely to be incorrect for such models [see Mahmassani et al. (2011)] and could introduce bias in the results, the consistency of the corresponding conclusions was verified within the study. In fact, after the analysis allows the noninfluential parameters to be individuated, the successive calibrations will provide direct feedback on the meaningfulness of the sensitivity analysis results. On the other hand, the independence of parameters is the basic assumption behind Equations (B.1) and (B.2)—and behind all the other formulas in the literature—that allow the calculation of the variances in the Sobol indexes and is otherwise unfeasible in cases of a number of nontrivial parameters (see Section 5.3.12).

Two simple distributions were initially tested for the two models: the triangular and the uniform. Such approach is customary when dealing with unknown pdfs (Law and Kelton, 2009) and in exploratory sensitivity analysis studies, in particular, where simple distributions are deemed sufficient for the scope of the analyses (Helton, 1993; Haan et al., 1998).

In defining input ranges whose proper assignment often exerts more influence on sensitivity analysis results than the knowledge of the actual pdfs (Haan et al., 1998; Helton, 1993), the first guess about the IDM parameters was based on the results of a study by Hoogendoorn and Hoogendoorn (2010). In that work, the distributions of five parameters (all except α and β) were estimated on four data sets of different sizes (1, 10, 25, and 100 trajectories). With the increase of size, they noted an improvement of the estimation results and a decrease of the input ranges. In this study, results from the sample of 25 trajectories were initially chosen as a fair compromise between the two tendencies. Corresponding upper and lower bounds for the IDM parameters are reported in the first two columns of Table B.1 as LB1 and UB1.

In the following section, however, it will be shown how results of a sensitivity analysis run on those ranges are not meaningful, suggesting that the ranges are not appropriate. Therefore, new ranges were defined and combining the information from the study by Hoogendoorn and Hoogendoorn with physical reasoning based on the actual characteristics of the trajectories used as fixed inputs (reported at the bottom of Figure B.1) and on the model features and meanings described in Section B.1.

For instance, the lower bound of the maximum speed was set at the value of the actual maximum speed observed in the data, while the upper one was set at a reasonably higher value. Choosing a lower bound value below the maximum speed observed in the data would have made the follower drift apart from his leader artificially. Conversely too high upper bound would have included fictitious amount of uncertainty. Wider intervals were adopted for $\ddot{S}_f$ and b_f, given the variety of meanings and implications on model behavior (see Section B.1), while a narrower interval chosen for ΔS^0 provided the experimental evidence.

In general, the principle was adopted to seek the best trade-off between decreasing ranges that may exclude likely driving behaviors and increasing

Table B.1 Input factor ranges adopted for IDM and the Gipps model for data sets 30B and 30C

Parameter	IDM						Gipps			
			Set 30B		Set 30C		Set 30B		Set 30C	
	LBl	UBl	LB	UB	LB	UB	LB	UB	LB	LB
α	0.1	10.0	1.0	8.0	1.0	8.0	—	—	—	—
β	0.1	10.0	1.0	5.0	1.0	5.0	—	—	—	—
$\ddot{S}_f$	0.5	2.5	1.0	8.0	1.0	8.0	4.7	8.0	3.4	8.0
b_f	0.0	2.5	0.5	7.5	0.5	7.5	2.0	8.0	2.0	8.0
ΔS^0	5.0	25.0	0.1	2.0	0.1	1.0	0.1	2.0	0.1	1.0
T	0.0	2.0	0.1	1.0	0.1	1.0	—	—	—	—
$\dot{S}_f$	20.0	60.0	18.4	25.0	14.1	25.0	18.4	25.0	14.1	25.0
γ	—	—			—	—	−4.0	4.0	−4.0	4.0
$\hat{b}$	—	—			—	—	2.0	8.0	2.0	8.0
Θ	—	—			—	—	0.05	0.5	0.05	0.5
τ	—	—			—	—	0.1	1.0	0.1	1.0

Note: First two columns of the IDM model were directly obtained by Hoogendoorn and Hoogendoorn (2010).

ranges that tend to encompass disproportionate amounts of uncertainty fictitiously, making model predictions of no practical use.

For the Gipps model, upper and lower bounds of parameter ranges were initially chosen based on the author's experience and successively refined through iterative tests and the use of scatter plots (in particular for parameter γ, which is usually kept fixed). The ranges finally adopted for both models are reported in Table B.1, for data sets 30B and 30C.

B.3 STEP C: PROPAGATION OF UNCERTAINTY SOURCES INTO OUTPUTS

The third step was propagating the uncertainty in the inputs into the outputs by the model. Propagation was carried out in a Monte Carlo framework following the method described in Section 5.3.12, in particular, through the generation of matrices [Equations (5.21), (5.22), and (5.23)] following a quasi-random sequence of numbers.

To evaluate the sensitivity indices for the seven input factors (parameters) of the IDM, $N \cdot (7 + 2)$ model runs were needed. The N was set at 10,000 after verification that the Sobol indices were stable for the value (see next section). With regards to the Gipps model, the amended version with eight parameters, given by Equations (B.3), (B.4), and (B.7) and having eight parameters was applied. $N \cdot (8 + 2)$ model evaluations were then performed.

B.4 STEP C: SENSITIVITY ANALYSIS AND RANKING

Once obtained model evaluations for the matrices [(5.21), (5.22), and (5.23)], first-order sensitivity indices and total effects through Equations (5.24) and (5.25) were calculated. In the following we show only results for the case of uniform distributions of the model parameters are shown because no particular qualitative and quantitative differences were observed for the triangular distribution case. Indices are presented for both speed and inter-vehicle spacing measures, for the two models.

B.4.1 Intelligent driver model

Results presented in Figure B.2 refer to the analysis carried out on the parameter ranges obtained from the study of Hoogendoorn and Hoogendoorn (2010) on the urban congested trajectory data set (30C). The total sensitivity index of each parameter is plotted against the number of model evaluations N for the speed (left plot) and for the spacing (right plot). It can be noted that the indices were stabilized as $N = 3,000$.

Results for both the measures were similar, with the maximum speed, $\dot{S}_f$ with all its interactions explaining about 80% of the entire output variance, followed by parameters α and β. All other parameters resulted not significant.

Figure B.3 presents results with the new parameter ranges. It is worth noting that the pattern is completely different from that in Figure B.2, both for speed and spacing. Specifically, the results confirm the guess that the initial ranges chosen were not appropriate. The high significance of the maximum speed shown in Figure B.3 is clearly the effect of incorrect definition of the input range for that parameter (and for the others) as detailed in Section B.2.

With the new ranges, all parameters turned out to be significant except for the desired inter-vehicle spacing at stop ΔS^0, which does not account for a significant portion of the output variance and thus is the only parameter for which calibration may be omitted. The most influencing parameter was the minimum time headway T for all the trajectories and measures considered. Parameters α and β accounted for a significant share of the output variance, confirming the need to include them in any model calibration and not considering them as constants.

Maximum acceleration $\ddot{S}$ explains a significant share of the speed variance (see Figures B.3a and c), while it exerts no influence on spacing (Figures B.3b and d). Conversely, maximum speed $\dot{S}$ accounts for a certain share of the spacing variance, while it is not influential on the speed

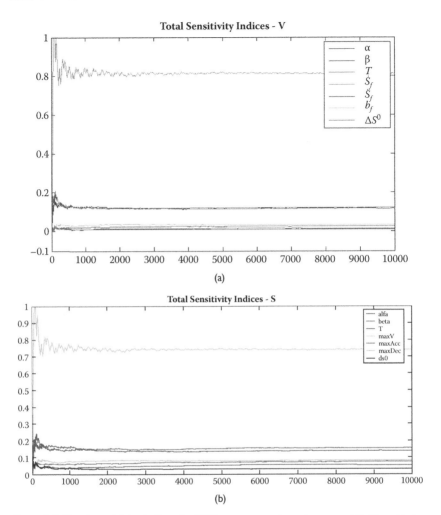

Figure B.2 Total sensitivity indices for IDM parameters based on speed (left) and inter-vehicle spacing (right) measures. Model is fed with the 30C leader's trajectory and with parameter values sampled from the uniform ranges defined by LBI and UBI in Table B.I. (*Source:* Ciuffo et al. 2014. With permission.)

(compare Figures B.3d and c). This highlights the concept that the choice of the parameters to calibrate may depend on the measure of performance adopted (in this case speed versus spacing).

Comparison of the left and right plots allows the impact of the two fixed inputs to be investigated. Indeed, in such a specific case, the parameter prioritization based on terms of the explained variance share does not change, although the share of variance explained by one parameter may vary significantly from one data set to the other. For example, maximum acceleration $\ddot{S}$

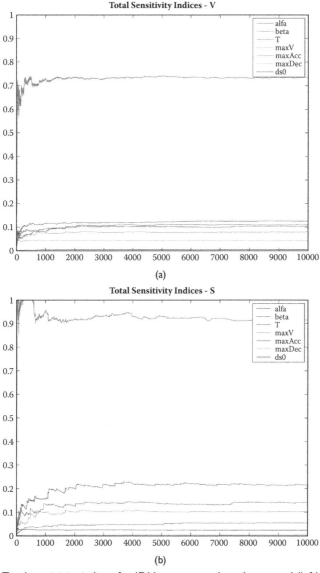

Figure B.3 Total sensitivity indices for IDM parameters based on speed (left) and inter-vehicle spacing (right) measures. Model is fed with the leader's trajectory from data sets 30C (upper plots) and 30B (lower plots) and with values of parameters sampled from uniform ranges defined by Columns 3 to 6 in Table B.I. Ciuffo, B., Punzo, V., and Montanino, M. 2012. The Calibration of Traffic Simulation Models: Report on the Assessment of Different Goodness-of-Fit Measures and Optimization Algorithms–COST Action TU0903 (MULTITUDE). JRC Scientific and Technical Reports, JRC 68403, Publications Office of the European Union, Luxembourg. *(continued)*

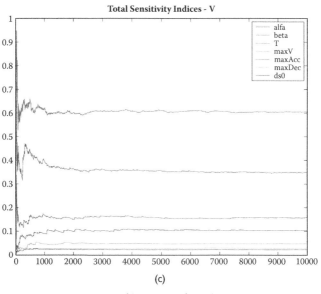

(c)

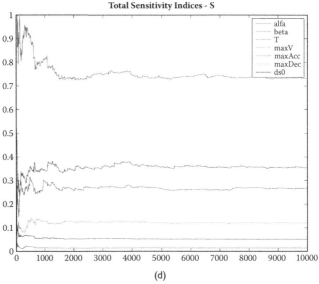

(d)

Figure B.3 (continued) Total sensitivity indices for IDM parameters based on speed (left) and inter-vehicle spacing (right) measures. Model is fed with the leader's trajectory from data sets 30C (upper plots) and 30B (lower plots) and with values of parameters sampled from uniform ranges defined by Columns 3 to 6 in Table B.I. Ciuffo, B., Punzo, V., and Montanino, M. 2012. The Calibration of Traffic Simulation Models: Report on the Assessment of Different Goodness-of-Fit Measures and Optimization Algorithms–COST Action TU0903 (MULTITUDE). JRC Scientific and Technical Reports, JRC 68403, Publications Office of the European Union, Luxembourg.

accounts for only 10% of the output variance in the urban data set (Figure B.1a) compared with 40% in the extra-urban data set (Figure B.1c). The impacts of these considerations on model calibration are further analyzed in Section B.2.

B.4.2 Gipps car-following model

In Figure B.4, results are presented for the urban trajectory 30C. Besides the total sensitivity indices (Figures B.4c and d), the first-order sensitivity indices (Figure B.4a and b) are shown for speed (left plots) and for spacing (right plots). In this case, all the results appear almost stable for N higher than 6,000.

It is possible to observe a slight difference between the results achieved using the spacing and those using the speed. In both the cases, the most important parameter is the follower's deceleration b_f. This was expected for a trajectory characterized by frequent and intense acceleration and deceleration phases. Other parameters explaining a considerable share of the output variance are the apparent reaction time τ and the follower's estimation of leader maximum deceleration $\hat{b}$.

It is interesting to notice the difference between the first-order and total sensitivity indices. The output variance explained by $\hat{b}$ is approximately constant, meaning that this parameter has negligible interactions with the others. Conversely, for b_f and τ, the share of variance explained in combination with other parameters (difference between total and first-order effects) has approximately the same magnitude. Their reciprocal interaction is therefore a good candidate to explain the most of such variance share, as expected from Equation (B.5).

Even more interesting are the results for the maximum speed $\dot{S}_f$, the maximum acceleration $\ddot{S}_f$, and parameter γ. Although they have negligible direct influence on the outputs (see Figure B.4a), they are influential when we consider interaction effects with other parameters (see Figure B.4c) with γ representing the second-most influential parameter. Again, this could be explained noting that these parameters all belong second ordered added terms added in Equation (B.7).

This clearly shows that first-order sensitivity indices alone are not sufficient to characterize the impacts of parameters on model outputs. In addition, with the Gipps model, we see the first evidence of the importance of calibrating parameter γ (that is, calibrating parameters α, β, and γ in the original formulation), which is usually considered a constant.

The behavior previously outlined is less evident if we consider the results of a sensitivity analysis on the extra-urban trajectory 30B. In this case, the prioritization of parameters is the same as for data set 30C except for the total sensitivity indices using speed as measure. The role played by the acceleration part of the model and therefore by parameters $\dot{S}_f$, $\ddot{S}_f$, and γ is less significant because the acceleration and deceleration phases are much less frequent and important (see Figure B.5).

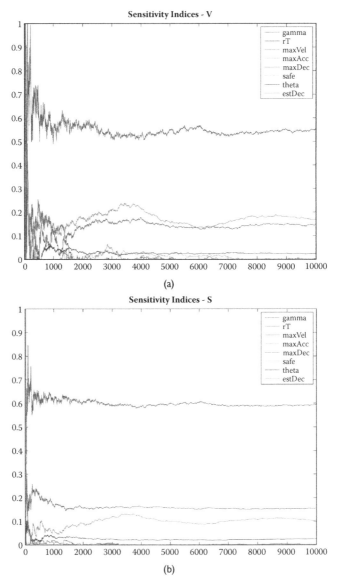

Figure B.4 First-order (a, b) and total (c, d) sensitivity indices for parameters of Gipps car-following model based on speed (left) and inter-vehicle spacing (right) measures. Model is fed with the leader's trajectory from data set 30C, with values of the parameters sampled from the uniform ranges in Table B.1. Ciuffo, B., Punzo, V., and Montanino, M. 2012. The Calibration of Traffic Simulation Models: Report on the Assessment of Different Goodness-of-Fit Measures and Optimization Algorithms–COST Action TU0903 (MULTITUDE). JRC Scientific and Technical Reports, JRC 68403, Publications Office of the European Union, Luxembourg. *(continued)*

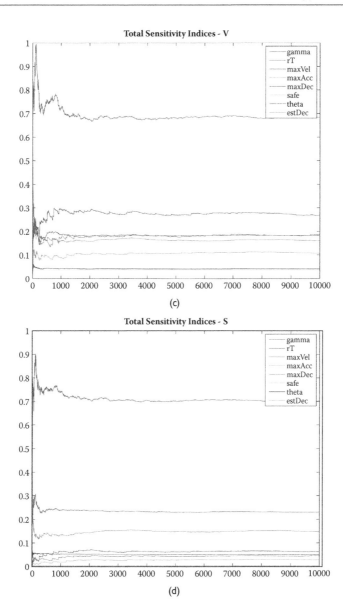

Figure B.4 (continued) First-order (a, b) and total (c, d) sensitivity indices for parameters of Gipps car-following model based on speed (left) and inter-vehicle spacing (right) measures. Model is fed with the leader's trajectory from data set 30C, with values of the parameters sampled from the uniform ranges in Table B.1. Ciuffo, B., Punzo, V., and Montanino, M. 2012. The Calibration of Traffic Simulation Models: Report on the Assessment of Different Goodness-of-Fit Measures and Optimization Algorithms—COST Action TU0903 (MULTITUDE). JRC Scientific and Technical Reports, JRC 68403, Publications Office of the European Union, Luxembourg.

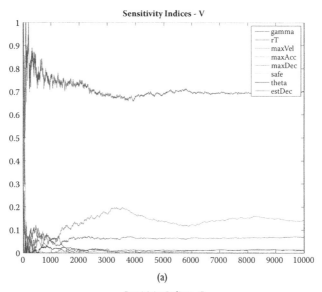

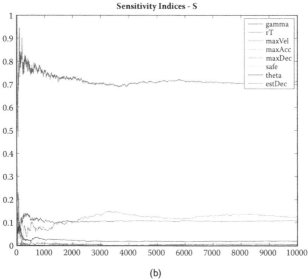

Figure B.5 First-order (a, b) and total (c, d) sensitivity indices for parameters of Gipps car-following model, based on speed (left) and inter-vehicle spacing (right) measures. Model is fed with the leader's trajectory from data set 30B, with values of the parameters sampled from the uniform ranges in Table B.1. Ciuffo, B., Punzo, V., and Montanino, M. 2012. The Calibration of Traffic Simulation Models: Report on the Assessment of Different Goodness-of-Fit Measures and Optimization Algorithms–COST Action TU0903 (MULTITUDE). JRC Scientific and Technical Reports, JRC 68403, Publications Office of the European Union, Luxembourg. (*continued*)

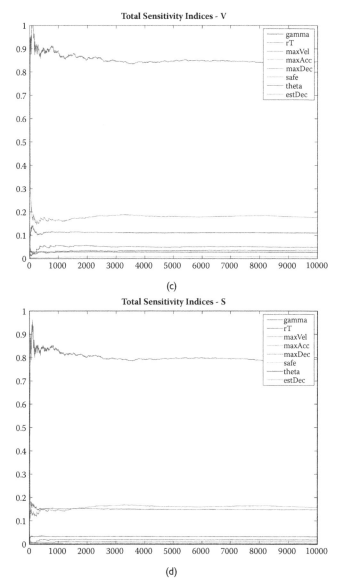

(c)

(d)

Figure B.5 (continued) First-order (a, b) and total (c, d) sensitivity indices for parameters of Gipps car-following model, based on speed (left) and inter-vehicle
spacing (right) measures. Model is fed with the leader's trajectory from data
set 30B, with values of the parameters sampled from the uniform ranges in
Table B.1. Ciuffo, B., Punzo, V., and Montanino, M. 2012. The Calibration
of Traffic Simulation Models: Report on the Assessment of Different
Goodness-of-Fit Measures and Optimization Algorithms–COST Action
TU0903 (MULTITUDE). JRC Scientific and Technical Reports, JRC 68403,
Publications Office of the European Union, Luxembourg.

B.5 IMPACT OF SENSITIVITY ANALYSIS ON MODEL CALIBRATION

In this section, the quess applying a variance-based technique to perform a sensitivity analysis of car-following models in a factor fixing setting is verified. Whether such a technique may be useful to the individuate subset of model parameters to calibrate without sensibly affecting model performance was investigated.

The two models were calibrated by hypothesizing alternative model configurations differing for the choices of the fixed and uncertain parameters. The different configurations were established on the basis of the prioritization of parameters provided by the sensitivity analysis study. In these calibration experiments, an increasing number of fixed parameters was considered to investigate the corresponding decrease in model ability to reproduce the actual trajectory. A synthetic description of the plan of experiments is reported in Table B.2.

The 40 experiments listed in the table were based on the total sensitivity indices of both the models parameters, for the combinations of trajectories (30B and 30C) and measures of performance (speed and spacing). In the table, each calibration experiment is represented by a unique code composed by three terms. The first term refers to the data set applied (B = 30B, C = 30C); the second term refers to the MoP (S = spacing and V = speed), while the third covers the number of uncertain parameters (i.e., to be calibrated). In particular, the *complete* term means considering all the model parameters uncertain (including the α, β, and γ constants). *Classic* indicates the standard versions of the models ($\beta = 2$ for the IDM and $\alpha = 2.5$, $\beta = 0.025$, and $\gamma = 0.5$ for the Gipps model).

B.5.1 Formulation and solution of calibration problem

Calibrating a simulation model consists of finding parameter values that allow the best possible fitting of the model to the behavior of a real system. It can be expressed in terms of the solution of a constrained minimization problem in which the objective function expresses the deviation of the simulated measurements from those observed:

$$\min_{\beta,\gamma} f(M^{\mathrm{obs}}, M^{\mathrm{sim}}) \tag{B.9}$$

possibly subject to the following constraints:

$$l_{\beta,i} \le \beta_i \le u_{\beta,i} \qquad i = 1, \ldots, m$$

$$l_{\gamma,j} \le \gamma_j \le u_{\gamma,j} \qquad j = 1, \ldots, n$$

Table B.2 Experimental calibration plan

	IDM			Gipps model	
ID	Experiment code	Parameters calibrated	ID	Experiment code	Parameters calibrated
I	B_S_ complete	All	19	B_S_complete	All
2	B_S_classic	All but α and β	20	B_S_classic	All but α, β and γ
3	B_S_6	$b_f, \Delta S^0, S_f, T, \alpha, \beta$	21	B_S_3	$b_f, \hat{b}, \tau$
4	B_S_4	S_f, T, α, β	22	B_S_I	b_f
5	B_S_3	T, α, β	23	B_S_I	$\hat{b}$
6	B_S_I	T	24	B_S_I	T
7	C_S_ complete	All	25	B_S_I	S_f
8	C_S_classic	All but α and β	26	C_S_complete	All
9	C_S_4	$\dot{S}_f, T, \alpha, \beta$	27	C_S_classic	All but α, β and γ
10	C_S_I	T	28	C_S_3	$b_f, \hat{b}, \tau$
11	B_V_ complete	All	29	C_S_I	b_f
12	B_V_classic	All but α and β	30	C_S_I	$\hat{b}$
13	B_V_4	$\ddot{S}_f, T, \alpha, \beta$	31	C_S_I	T
14	B_V_2	$\ddot{S}_f, T$	32	C_S_I	S_f
15	C_V_ complete	All	33	B_V_complete	
16	C_V_classic	All but α and β	34	B_V_classic	All but α, β and γ
17	C_V_5	$S_f, S_f, T, \alpha, \beta$	35	B_V_3	$b_f, \hat{b}, \tau$
18	C_V_I	T	36	B_V_I	b_f
			37	C_V_complete	All
			38	C_V_classic	All but α, β and γ
			39	C_V_5	$\gamma, b_f, \hat{b}, \tau, S_f$
			40	C_V_I	b_f

The experiment code comprises three terms. The first refers to the data set applied (B = 30B, C = 30C); the second to the MoP (S = spacing, V = speed), and the third to the number of uncertain parameters. *Complete* means considering all model parameters uncertain. *Classic* indicates standard versions of models.

and potentially also to other constraints:

$$g_k(\beta_i, \gamma_j)?b_k \qquad i = 1...m, \; j = 1...n, \; k = 1...l$$

where β_i and γ_i are, respectively, the vectors of continuous and discrete model parameters potentially belonging to m different classes of simulation subjects; $f(\cdot)$ is the objective function (or fitness or loss function) to be minimized and measures the distance between the simulated and observed traffic measurements, M^{sim} and M^{obs}; $l_{\beta,i}$, $l_{\gamma,i}$, $u_{\beta,i}$, $u_{\gamma,i}$, are model parameter lower and upper bounds; $g_k(\beta_i, \gamma_j)$ is a scalar valued linear or nonlinear function of model parameters β_i, γ_j, that calculates the left side of the k-th constraint; b_k is a constant value equal to the right side of the k-th constraint; and ? is one three relational operators: $\leq$, $\geq$, or $=$.

The problem in Equation (B.9) cannot be solved analytically because we are dealing with a simulation model. An optimization algorithm is used instead. In the present case, we used the Opt/Quest Multistart algorithm as implemented in the Lindo API optimization package (Lindo, 2003). The good performance of this algorithm in dealing with similar problems was tested in several studies [Ciuffo et al. (2008); Ciuffo et al. (2011)].

In the present study, M^{sim} and M^{obs} are, respectively, the simulated and observed trajectory data. A trajectory is defined as the time series of the positions or speeds assumed by a vehicle over its path. As a measure of goodness of fit of simulated and observed trajectories, the root mean squared error ($RMSE$) was applied, following the results of the extensive study carried out in Punzo et al. (2011).

B.5.2 Results

Calibration results are reported in Table B.3 for the IDM and Table B.4 for the Gipps model. The tables show that the number of algorithm iterations considerably decreases with decreases in the numbers of parameters calibrated, i.e., with the less sensitive parameters considered fixed while the value of the objective function does not increase sharply. This suggests a positive trade-off between a decrease in computational cost and result accuracy.

This feature could be particularly significant for computationally expensive calibrations, that is, if the number of parameters exceeds 10 or 15. For instance, in the experiment with ID = 5, in which the IDM was been calibrated for only three out of seven parameters (T, α, and β), the results for spacing and speed are not considerably worse than those of the calibrations carried out with more parameters. This is particularly relevant for the Gipps model. Indeed, if we compare calibration 23 (B_S_1) to 20 (B_S_classic), we observe approximately the same result in terms of the objective function with three times fewer iterations and only one parameter calibrated against the seven of the classical version.

Table B.3 IDM calibration results

ID	Code	Iterations	ObjV	ObjS	$\ddot{S}_f$	b_f	ΔS^0	$\dot{S}_f$	T	α	β
1	B_S_complete	5,030	0.509	1.416	2.06	7.50	1.94	18.90	0.55	8.00	5.00
2	B_S_classic	3,464	0.501	1.421	2.83	2.92	2.00	20.58	0.54	8.00	2.00
3	B_S_6	2,338	0.506	1.431	4.73	1.00	2.00	22.40	0.54	8.00	1.00
4	B_S_4	2,214	0.503	1.487	4.73	1.00	1.31	23.33	0.61	8.00	1.00
5	B_S_3	1,864	0.562	1.610	4.73	1.00	1.31	18.37	0.59	8.00	5.00
6	B_S_1	1,556	0.583	2.343	4.73	1.00	1.31	18.37	0.42	4.00	2.00
7	C_S_complete	4,228	0.542	1.469	1.41	0.84	0.61	18.26	0.49	1.00	5.00
8	C_S_classic	5,287	0.524	1.539	1.41	0.50	0.25	18.44	0.40	1.00	2.00
9	C_S_4	1,991	0.643	2.207	3.06	1.00	0.19	14.13	0.52	1.00	5.00
10	C_S_1	1,598	0.710	2.878	3.06	1.00	0.19	14.13	0.61	4.00	2.00
11	B_V_complete	3,974	0.445	3.808	3.26	7.50	2.00	21.82	0.86	8.00	1.95
12	B_V_classic	3,732	0.445	3.799	3.17	7.50	2.00	21.73	0.86	8.00	2.00
13	B_V_4	2,114	0.501	3.315	2.02	1.00	1.31	18.37	0.74	8.00	1.95
14	B_V_2	1,504	0.518	3.509	2.03	1.00	1.31	18.37	0.62	4.00	2.00
15	C_V_complete	3,916	0.448	1.803	1.58	7.50	0.81	18.37	0.54	1.00	4.18
16	C_V_classic	3,652	0.458	1.909	1.57	3.79	0.24	18.97	0.40	1.00	2.00
17	C_V_5	3,047	0.482	1.822	1.47	1.00	0.19	19.34	0.37	1.00	1.40
18	C_V_1	1,344	0.683	3.653	3.06	1.00	0.19	14.13	0.86	4.00	2.00

Calibration identifications refer to Table B.2. Calibrated parameters are shown in italics. Other parameters are kept fixed to randomly generated values. Measure used for calibration is shown in bold. Red lines indicate calibration using trajectory 30C, while the others use trajectory 30B. In calibrations with ID from 1 to 10, spacing was adopted as MoP. The others used the speed. Calibrations with ID from 1 to 6 and 11 to 14 were performed on trajectory 30C while the others on trajectory 30B.

Table B.4 Gipps model calibration results

ID	Code	Iterations	ObjV	Obj S	α	β	γ	$\ddot{S}_f$	ΔS^0	b_f	$\hat{b}$	θ	τ	$\dot{S}_f$
19	B_S_complete	14,012	0.533	1.363	1.00	1.00	−2.90	4.93	1.99	−4.38	−3.88	0.50	0.20	21.71
20	B_S_classic	21,767	0.524	1.515	2.50	0.03	0.50	5.59	1.67	−3.34	−3.38	0.11	0.52	21.98
21	B_S_3	16,439	0.531	1.549	1.00	0.97	−0.75	6.47	1.23	−3.50	−3.45	0.32	0.41	21.89
22	B_S_1	4,743	0.527	1.636	1.00	0.97	−075	6.47	1.23	−6.10	−5.56	0.29	0.48	21.48
23	B_S_1	5,594	0.529	1.590	1.00	0.97	−0.75	6.47	1.23	−5.56	−5.14	0.29	0.48	21.48
24	B_S_1	1,837	0.552	3.177	1.00	0.97	−0.75	6.47	1.23	−5.56	−5.19	0.23	0.55	21.48
25	B_S_1	3,931	0.657	7.128	1.00	0.97	−0.75	6.20	1.11	−5.56	−5.56	0.29	0.48	21.48
26	C_S_complete	27,889	0.553	1.490	1.00	1.00	−3.75	3.36	0.84	−8.00	−7.16	0.22	0.45	25.00
27	C_S_classic	16,157	0.678	2.779	2.50	0.03	0.50	3.36	0.43	−3.55	−4.25	0.11	0.70	14.13
28	C_S_3	15,207	0.696	2.546	1.00	0.97	−0.75	5.83	0.63	−3.87	−5.00	0.32	0.36	19.90
29	C_S_1	5,774	0.677	2.525	1.00	0.97	−0.75	5.83	0.63	−5.79	−5.56	0.29	0.48	19.23
30	C_S_1	7,465	0.682	2.502	1.00	0.97	−0.75	5.83	0.63	−5.56	−5.31	0.29	0.48	19.23
31	C_S_1	1,761	0.665	2.623	1.00	0.97	−0.75	5.83	0.63	−5.56	−5.19	0.23	0.55	19.23
32	C_S_1	5,664	0.810	4.721	1.00	0.97	−0.75	4.35	0.58	−5.56	−5.56	0.29	0.48	19.23
33	B_V_complete	12,949	0.444	4.287	1.00	0.99	−2.25	4.78	2.00	−5.19	−4.37	0.18	0.95	21.86
34	B_V_classic	7,832	0.472	4.292	2.50	0.03	0.50	6.91	2.00	−5.61	−5.27	0.07	0.96	20.58
35	B_V_3	8,221	0.478	3.860	1.00	0.97	−075	6.47	1.23	−7.13	−6.37	0.32	0.77	21.89
36	B_V_1	6,365	0.504	3.134	1.00	0.97	−0.75	6.47	1.23	−5.05	−5.56	0.29	0.48	21.48
37	C_V_complete	12,879	0.493	2.107	1.00	1.00	−3.32	3.50	0.58	−5.71	−5.50	0.09	0.81	25.00
38	C_V_classic	5,846	0.622	4.317	2.50	0.03	0.50	3.59	0.74	−5.75	−7.55	0.05	0.90	19.60
39	C_V_5	3,942	0.423	2.262	1.00	1.00	−0.58	1.34	0.81	−7.64	−7.77	0.29	0.13	20.60
40	C_V_1	5,162	0.639	3.323	1.00	0.97	−0.75	5.83	0.63	−4.54	−5.56	0.29	0.48	19.23

Calibration identifications refer to Table B.2. Calibrated parameters shown in italic. Other parameters are kept fixed to randomly generated values. Measures used for calibration are shown in bold. Red lines refer to calibration using trajectory 30C, while the others use trajectory 30B. In calibrations with ID from 1 to 10, spacing was adopted as MoP. The others used the speed. Calibrations with ID from 1 to 6 and 11 to 14 were performed on trajectory 30C while the others on trajectory 30B.

Such trade-offs are not always so positive. For example, in ID = 6 (B_S_1) in which only one parameter (minimum time headway) of the IDM is calibrated, the results are worse than those for the corresponding Gipps model experiment ID = 22 (B_S_1). However, this could be somewhat expected because minimum time headway accounts for 40% of the IDM output variance in terms of main effect and for 70% when considering the interaction effects with other parameters, while the follower's deceleration b_f in the corresponding setting for the Gipps model accounts, respectively, for 73 and 79% of the output variance. Similar scarce performances hold also for calibration 10 (C_S_1) in which minimum time headway accounts for 90% of the variance considering the interaction effects, but only 70% as a main effect.

Other evidence of the meaningfulness of sensitivity analysis prioritization can be found in the results of calibrations 25 and 32. Only one parameter of the Gipps model was calibrated, as in experiments 22 and 29. However, unlike those calibrations in which the model was calibrated against the most sensitive parameter (b_f) and the results are more than satisfactory, the model in experiments 25 and 32 was calibrated against the scarcely sensitive parameter S double pointed and the calibration performances were unsatisfactory. However, we must note that this also depends on how distant the sensitive parameters are from their "best" values, but the results achieved highlight this point.

Furthermore, for both the IDM and the Gipps model, the calibration of the parameters usually considered fixed leads to an overall improvement of the objective function. This is more evident in the Gipps model, where the results clearly confirm the outputs of the sensitivity analysis.

Comparing instead the results of calibrations using speed and spacing, we observe that calibrating on spacing provides a better fit also to the speed profile, while the opposite is not true. This was already observed by Punzo and Simonelli (2005) and again suggests that it is preferable to calibrate car-following models using the time series of spacing instead of speed. Finally, we note that the calibration results are consistent with the expectations despite the parameter independence issue.

B.6 CONCLUSION AND RECOMMENDATIONS

In the application presented, variance-based techniques for model sensitivity analysis were discussed and applied to two car-following models. Throughout the case, it is argued that the application of such methods is crucial for true comprehension and correct use of these models, specifically the issues of model parameter estimation and calibration.

The concept is supported by the outcomes of the technique applied and by the calibrations of the two car-following models using only the parameters suggested by the sensitivity analysis. Important issues arising during set-up of a sensitivity analysis were investigated and discussed.

First, the importance of the data assimilation phase was highlighted by the presentation of the controversial results in Figure B.2 caused by an incorrect or inconsistent definition of the input space. The results reveal the effect of cutting off some of the input uncertainty by considering too-tight boundaries for the parameters. This has notable implications also for model calibration, where the smallest possible searching space for the optimization algorithms is generally sought.

The recommendation is therefore to obtain prior information about input distribution, possibly from data sets covering a wide spectrum of traffic patterns (e.g., through preliminary parameter estimation). An alternative would be to perform a number of preliminary tests (sensitivity analyses) to find the right balance.

Figure B.4 and Figure B.5 depict the influence on the results of the information used to feed the models. The use of trajectories containing different dynamics (urban or extra-urban) allowed us to quantify changes in the relative importances of model parameters in affecting the output variances of different traffic patterns. We showed that the application of such techniques may allow the richness of a particular data set to be also investigated if a priori information on the sensitivity indices of the model parameters is available, e.g., calculated over wider datasets.

Comparison of first-order sensitivity indices and total effects of a parameter tell us whether that parameter has higher-order effects on model outputs. Such results, and the ranking of parameters they generate, provide crucial information for model calibration too, as shown in the cases of the IDM and Gipps model in which some parameters, generally considered fixed, were demonstrated to account for a not-negligible share of output uncertainty. This suggests that those parameters should be calibrated in future applications of such models.

The sensitivity analysis allowed us also to evaluate the parsimony of the two models, that is, the ability to describe reality with a minimum of adjusting parameters (Young et al., 1996). Neither model was totally parsimonious—with one parameter showing higher relative importance than others. In general, however, we can consider the IDM more parsimonious, with all the parameters but one explaining a significant share of the output variance.

Results with the Gipps model, on the other hand, may require further investigations, depending on the problem of input correlation introduced by Equation (B.6). While such correlation is not expected to be strong if as introduced by a structural equation of the model and if the rejection approach adopted is correct in principle, further research is needed to ascertain the correctness of results. Calibration results, however, seemed to confirm the reliability of the results achieved.

Index